IN-SITU DEEP SOIL IMPROVEMENT

Proceedings of sessions sponsored by the
Geotechnical Engineering Division of the
American Society of Civil Engineers
in conjunction with the
ASCE National Convention in Atlanta, Georgia,
October 9-13, 1994

Geotechnical Special Publication No. 45

Edited by Kyle M. Rollins

Published by the
American Society of Civil Engineers
345 East 47th Street
New York, New York 10017-2398

ABSTRACT

This proceedings, *In-Situ Deep Soil Improvement*, contains papers presented at sessions sponsored by the Geotechnical Engineering Division of ASCE in conjunction with the ASCE National Convention in Atlanta, Georgia, October 9-13, 1994. They mainly discuss the use of deep dynamic compaction for densification of collapsible soils, liquefiable soils and waste materials, and provide a practical summary of what has been learned regarding the potential for improvement in these materials. In addition to the topic of dynamic compaction, other methods for in-situ soil improvement, such as stone columns, deep soil mixing, and vacuum consolidation, are presented.

Library of Congress Cataloging-in-Publication Data

In-situ deep soil improvement: proceedings of sessions sponsored by the Geotechnical Engineering Division of the American Society of Civil Engineers in conjunction with the ASCE National Convention in Atlanta, Georgia, October 9-13, 1994 / edited by Kyle M. Rollins.
   p. cm.
   Includes index.
   ISBN 0-7844-0058-X
   1. Soil stabilization—Congresses. 2. Soil consolidation-Congresses. 3. Soil stabilization—Case studies—Congresses. 4. Soil consolidation—Case studies-Congresses. I. Rollins, Kyle M. II. American Society of Civil Engineers. Geotechnical Engineering Division. III. ASCE National Convention (1994: Atlanta, Ga.) IV. Series.
TA710.A1I388 1994   94-34442
624.1'5136—dc20   CIP

The Society is not responsible for any statements made or opinions expressed in its publications.

Photocopies. Authorization to photocopy material for internal or personal use under circumstances not falling within the fair use provisions of the Copyright Act is granted by ASCE to libraries and other users registered with the Copyright Clearance Center (CCC) Transactional Reporting Service, provided that the base fee of $2.00 per article plus $.25 per page copied is paid directly to CCC, 222 Rosewood, Drive, Danvers, MA 01923. The identification for ASCE Books is 0-7844-0058-X/94 $2.00 + $.25. Requests for special permission or bulk copying should be addressed to Permissions & Copyright Dept., ASCE.

Copyright © 1994 by the American Society of Civil Engineers, All Rights Reserved.
Library of Congress Catalog Card No: 94-34442
ISBN 0-7844-0058-X
Manufactured in the United States of America.

PREFACE

This special technical publication constitutes the proceedings of two sessions at the ASCE National Convention held in Atlanta, Georgia in October 1994. The two technical sessions entitled, "In-Situ Ground Improvement by Deep Dynamic Compaction" and "In-Situ Ground Improvement Case Histories" were sponsored by the Soil Improvement and Geosynthetics Committee of ASCE. The objective of the sessions is to share information gained from recent experience involving in-situ soil improvement techniques with practicing engineers, researchers and specialty contractors.

Deep dynamic compaction is becoming a more common method for densifying both solid waste and problem soils. Recently, a number of projects have been undertaken involving dynamic compaction for densification of collapsible soils, liquefiable soils and waste materials. The papers contained in this publication provide a practical summary of what has been learned regarding the potential for improvement in these materials and the suitability of various methods for evaluating improvement. In addition to case histories on dynamic compaction, two state-of-the-practice are presented. The first describes U.S. experience with dynamic compaction for collapsible soils and the second relates Bureau of Reclamation experience with dynamic compaction in treating liquefiable foundations soils below earth dams. Several additional papers describe other methods being used for in-situ soil improvement. These papers present case histories which illustrate the application of methods such as stone columns, deep soil mixing and vacuum consolidation.

It is the current practice of the Geotechnical Engineering Division that each paper published in a special publication be reviewed for content and quality. Reviews of papers for publication in this volume were conducted by members of the Soil Improvement and Geosynthetics committee and by other ASCE members with expertise in the subject areas. Each paper included in this volume has received at least two positive peer reviews and authors were given the opportunity to modify their papers based on reviewers suggestions prior to final submittal of the papers. All papers published are eligible for discussion in the *Journal of Geotechnical Engineering* and they are also eligible for ASCE awards.

The editor of this volume expresses appreciation to all the reviewers and authors who made the publication of this proceedings possible and to Shiela Menaker who provided needed assistance in a timely and efficient manner.

Kyle M. Rollins, M. ASCE
Assoc. Prof. of Civil Engineering
Brigham Young University

CONTENTS

A. In-Situ Ground Improvement by Deep Dynamic Compaction

Dynamic Compaction to Remediate Liquefiable Embankment Foundation Soils
  Karl Dise, Michael G. Stevens, and J. Lawrence Von Thun 1
U.S. Experience with Dynamic Compaction of Collapsible Soils
  Kyle M. Rollins and Ji-Hyoung Kim 26
Evaluation of Dynamic Compaction of Low Level Waste Burial Trenches
Containing B-25 Boxes
  Scott R. McMullin 44
Dynamic Compaction: Two Case Histories Utilizing Innovative Techniques
  Albert A. Bayuk and Andrew D. Walker 55
Dynamic Compaction Used as a Winter Construction Expedient
  Lawrence F. Johnsen and Christopher Tonzi 68

B. In-Situ Ground Improvement Case Histories

Dynamic Compaction of Saturated Silt and Silty Sand—A Case History
  Jean C. Dumas, Nelson F. Beaton, and Jean-François Morel 80
Prospects of Vacuum-Assisted Consolidation for Ground Improvement of Coastal
and Offshore Fills
  S. Thevanayagam, E. Kavazanjian, A. Jacob, and I. Juran 90
Soil Mix Walls in Difficult Ground
  David S. Yang and Shigeru Takeshima 106
Control of Settlement and Uplift of Structures Using Short Aggregate Piers
  Evert C. Lawton, Nathaniel S. Fox and Richard L. Handy 121
A Deep Soil Mix Cutoff Wall at Lockington Dam, Ohio
  Andrew D. Walker 133
Soil Improvement for Prison Facility
  Frances B. Gularte and S. Stuart Williams, Jr. .... *

Subject Index 147

Author Index 149

---

*Manuscript not available at time of printing.

GEOTECHNICAL SPECIAL PUBLICATIONS

1) TERZAGHI LECTURES
2) GEOTECHNICAL ASPECTS OF STIFF AND HARD CLAYS
3) LANDSLIDE DAMS: PROCESSES RISK, AND MITIGATION
4) TIEBACKS FOR BULKHEADS
5) SETTLEMENT OF SHALLOW FOUNDATION ON COHESIONLESS SOILS: DESIGN AND PERFORMANCE
6) USE OF IN SITU TESTS IN GEOTECHNICAL ENGINEERING
7) TIMBER BULKHEADS
8) FOUNDATIONS FOR TRANSMISSION LINE TOWERS
9) FOUNDATIONS AND EXCAVATIONS IN DECOMPOSED ROCK OF THE PIEDMONT PROVINCE
10) ENGINEERING ASPECTS OF SOIL EROSION DISPERSIVE CLAYS AND LOESS
11) DYNAMIC RESPONSE OF PILE FOUNDATIONS— EXPERIMENT, ANALYSIS AND OBSERVATION
12) SOIL IMPROVEMENT - A TEN YEAR UPDATE
13) GEOTECHNICAL PRACTICE FOR SOLID WASTE DISPOSAL '87
14) GEOTECHNICAL ASPECTS OF KARST TERRIANS
15) MEASURED PERFORMANCE SHALLOW FOUNDATIONS
16) SPECIAL TOPICS IN FOUNDATIONS
17) SOIL PROPERTIES EVALUATION FROM CENTRIFUGAL MODELS
18) GEOSYNTHETICS FOR SOIL IMPROVEMENT
19) MINE INDUCED SUBSIDENCE: EFFECTS ON ENGINEERED STRUCTURES
20) EARTHQUAKE ENGINEERING & SOIL DYNAMICS (II)
21) HYDRAULIC FILL STRUCTURES
22) FOUNDATION ENGINEERING
23) PREDICTED AND OBSERVED AXIAL BEHAVIOR OF PILES
24) RESILIENT MODULI OF SOILS: LABORATORY CONDITIONS
25) DESIGN AND PERFORMANCE OF EARTH RETAINING STRUCTURES
26) WASTE CONTAINMENT SYSTEMS; CONSTRUCTION, REGULATION, AND PERFORMANCE
27) GEOTECHNICAL ENGINEERING CONGRESS
28) DETECTION OF AND CONSTRUCTION AT THE SOIL/ROCK INTERFACE
29) RECENT ADVANCES IN INSTRUMENTATION, DATA ACQUISITION AND TESTING IN SOIL DYNAMICS
30) GROUTING, SOIL IMPROVEMENT AND GEOSYNTHETICS
31) STABILITY AND PERFORMANCE OF SLOPES AND EMBANKMENTS II (A 25-YEAR PERSPECTIVE)
32) EMBANKMENT DAMS-JAMES L. SHERARD CONTRIBUTIONS
33) EXCAVATION AND SUPPORT FOR THE URBAN INFRASTRUCTURE
34) PILES UNDER DYNAMIC LOADS
35) GEOTECHNICAL PRACTICE IN DAM REHABILITATION
36) FLY ASH FOR SOIL IMPROVEMENT
37) ADVANCES IN SITE CHARACTERIZATION: DATA ACQUISITION, DATA MANAGEMENT AND DATA INTERPRETATION
38) DESIGN AND PERFORMANCE OF DEEP FOUNDATIONS: PILES AND PIERS IN SOIL AND SOFT ROCK
39) UNSATURATED SOILS
40) VERTICAL AND HORIZONTAL DEFORMATIONS OF FOUNDATIONS AND EMBANKMENTS
41) PREDICTED AND MEASURED BEHAVIOR OF FIVE SPREAD FOOTINGS ON SAND
42) SERVICEABILITY OF EARTH RETAINING STRUCTURES
43) FRACTURE MECHANICS APPLIED TO GEOTECHNICAL ENGINEERING
44) GROUND FAILURES UNDER SEISMIC CONDITIONS
45) IN-SITU DEEP SOIL IMPROVEMENT

DYNAMIC COMPACTION TO REMEDIATE LIQUEFIABLE EMBANKMENT FOUNDATION SOILS

[1]Karl Dise
[2]Michael G. Stevens
[3]J. Lawrence Von Thun

ABSTRACT

The Bureau of Reclamation (Reclamation) used deep dynamic compaction to remediate liquefiable soils and improve the seismic stability of a number of embankment dams. Its use at Jackson Lake Dam Modification, Stages I and II, Mormon Island Auxiliary Dam Modification, Phase I, and Steinaker Dam Modification, resulted in an evolution of our understanding of how best to use deep dynamic compaction to remediate both fine and course-grained liquefiable soils. Thus, experiences and lessons learned from the Jackson Lake and Mormon Island projects were used to enhance the procedures for the Steinaker Dam Modification, and culminated in the effective remedial treatment of difficult to treat silty-sands and sandy-silts. Presented are summaries of the dynamic compaction programs for each project, an overview of verification testing results and methodologies, and recommendations for applying dynamic compaction to remediate liquefiable soils.

1. JACKSON LAKE DAM MODIFICATIONS, STAGES I AND II

1.1 Project Description

Jackson Lake Dam is located in the Grand Teton National Park north of Jackson, Wyoming, USA. Engineering and geologic investigations revealed that the dam was subject to failure due to earthquake loading from large, nearby seismic sources. They also showed that a

---

[1] Civil/Geotechnical Engineer, Principle Designer, U.S. Bureau of Reclamation, Denver, CO, 80225.
[2] Civil/Geotechnical Engineer, U.S. Bureau of Reclamation, Denver, CO, 80225
[3] Chief, Concrete Dams Branch, U.S. Bureau of Reclamation, Denver, CO, 80225.

considerable portion of the embankment and its foundation were susceptible to liquefaction and required remediation. Structural modifications were specified to accomplish remediation objectives under two construction contracts, Stages I and II, completed in 1986 and 1988, respectively.

Removing the existing northern embankment and replacing it with a modern zoned earthfill structure addressed the seismic concerns of the 80-year old hydraulic fill. Foundation treatment requirements for the new embankment were determined based on extensive preconstruction geologic investigations and post-earthquake stability analyses of the dam and foundation. Deep dynamic compaction was selected to accomplish shallow (9 to 12-meter deep) foundation treatment for both Stages.

Stage I dynamic compaction was performed along a 700-meter length of foundation beneath the new embankment. Treatment extended to about 6 meters beyond the upstream and downstream toes of the new embankment. In Stage II, Reclamation specified Soil-Mix-Wall (SMW) remediation technology to perform deep remedial treatment work along 500 meters of the upstream and downstream new embankment toes, while dynamic compaction was used in the central portion for shallow treatment. A general layout of the Stage I and II remediation work is provided on Figures 1 and 2. The dynamic compaction areas for both Stages were broken up into individual approval and testing areas, defined as Sectors A through Q.

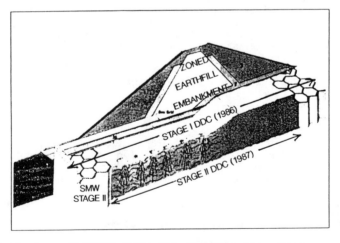

Figure 1. - Jackson Lake Dam Modification Features
(After Farrar, Wirkus, McLain, 1990)

The primary tool used to assess treatment effectiveness for both compaction programs was the Standard Penetration Test (SPT). SPT's were performed before construction for general site characterization and to provide assessments of liquefaction potential. Just prior to compaction, SPTs were performed to further define local site characteristics for individual approval areas (sectors) and to provide more detailed information across the entire foundation for pretreatment baseline information. SPTs were then performed after various phases within each compaction program, and subsequent to compaction, not only to evaluate the effectiveness of the specified program, but to better understand the dynamic compaction process.

Jackson Lake Dam Modification

Stage I Dynamic Compaction Plan

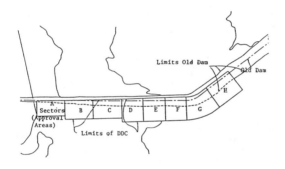

Jackson Lake Dam Modification

Stage II Dynamic Compaction Plan

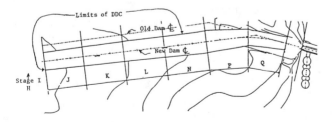

Figure 2. - Jackson Lake Dam Modifications - Stages I&II (After Farrar and Yasuda, 1991)

1.2 Site Geology

Site investigations indicated that a complex interbedded fluvial/lacustrine deposit would comprise the foundation beneath the new embankment (Farrar, Wirkus, McLain, 1990). The stratigraphy consisted of a heterogeneous mix of lenses of gravel, sand, and silt, generally grading from coarser to finer with depth. The deposits retained their character for distance of 120 meters or more. Preconstruction investigations indicated the presence of potentially liquefiable sands, silty sands and low plasticity silts to depths of up to 30 meters before encountering higher plasticity non-liquefiable lacustrine deposits. Although some fine-grained deposits were encountered within zones requiring remediation, the deposits appeared generally cohesionless and potentially suitable for treatment using deep dynamic compaction.

1.3 Compaction Programs

The compaction programs used for both Stages, and summarized in Table 1, were very similar. The major differences between the two programs were the Phase 1 "working surface preparation phase" (Luebke, et. al., 1985), the size and shape of the weight used for the Phase 5 "ironing" phase, and the lifting devices used to perform the work. Heavy tamping, Phases 2, 3 and 4 of each Stage, was performed with a tethered, 32 ton weight.

Attempts to drop the 30 ton weight in free-fall from available equipment during Stage I were unsuccessful. The contractor elected to use a single-part line to lift and drop the 32-ton weight from approximately 33 meters to achieve the energy requirements per drop that theoretically would be achieved with the specified 30-ton weight dropped 30 meters. The contractor used a laser measurement system to verify energy input per drop. This verification method was also adopted for work in Stage II, the Mormon Island Auxiliary Dam and Steinaker Dam modifications.

Twenty (20), fifteen (15) and (10) drops were applied to primary, secondary and tertiary impact locations for Phases 2, 3 and 4, respectively. A primary impact point spacing of 12 meters was selected to treat the majority of the foundation for both Stages.

Wick drains were specified in both Stages in areas where pre-compaction SPT data indicated the presence of fine grained liquefiable materials that would be difficult to treat, or where silts shielded deeper materials that required treatment. The wick drains were specified because they would enhance the dynamic compaction process by preventing the build-up of detrimental excess pore

TABLE 1. – COMPARISON OF COMPACTION PROGRAMS
USBR SAFETY OF DAMS PROJECTS (1986-1994)

Project (Year)	Phase	Compaction Programs Impact Location	Actual Drop Height (m)	Specified Drop Height (m)*	No. of Drops	Types of Equipment Lifting Device	Shape	Tamper Weight (tons)	Specified Weight (tons)*	Primary Impact Location Spacing (m)	Pore Pressure and Water Control Features Wick Drains Type	Depths	Dewatering Features	Types of Verification Testing	Verification Testing Locations	Treatment Area (m²)	Approximate Cost $/m² treated ground	Types of Materials Treated
Jackson Lake Dam Modification Stage I (1986)	1A	All	--	15	2	Single-line modified cranes***	Square	--	50	9 and 12	Alsdrain	9 to 12	Complete Perimeter Dewatering Trench 4.5-6 m deep	SPT (Primary) CPT, DMTs	Quaternary Drop Points	46,000	4.60	ML, SM SP-SM, SP, GM, GP
	1B	Primary	--	30	10		Octagonal	12	50									
	1C	Secondary	30+	15	2		.	12	50									
	2	Primary	30+	30	20		Square	12	50									
	3	Secondary	30+	30	15		.	12	50									
	4	Tertiary	30+	30	10		.	20	20									
	5	Entire Area	12	12	1-2													
Jackson Lake Dam Modification Stage II (1987)	1	All	--	12	4	Single-line dynamic compactors	Octagonal	--	20	12	Alsdrain	9 to 14	Complete Perimeter Dewatering Trench 4.5-6 m deep	SPT (Primary) X-hole x-wave, SASW, DMT, Stepped blade, Piezometers Undisturbed Sampling Seismic Tomography	Quaternary Drop Points	21,000	5.90	ML, SM SP-SM, SP, GM, GP
	2	Secondary	30+	30	30		.	12	50									
	3	Primary	30+	30	15		.	12	50									
	4	Tertiary	30+	30	10		.	12	50									
	5	Entire Area	9.5	12	2		.	20	20									
Mormon Island Auxiliary Dam Modification (1990)	1	All	--	9	4	Single-line dynamic compactors	Circular	--	55	15	None	None	Partial (2 sides) Perimeter Dewatering Trench 1.5-3 m deep	BPT (Primary) SPT, X-hole S-wave, SASW	All drop point locations	10,000	7.20	Dredged Tailings of: GP, GM, SP, SMs Some SM-MLs
	2	Primary	33+	30	30		.	55	55									
	3	Secondary	33+	30	15		.	55	55									
	4	Tertiary	33+	30	2		.	55	55									
	5	Entire Area	18	9														
Steinaker Dam Modification (1993)	1	All	18	12-18	2	Single-line dynamic compactors	Circular	30	30	15	Ameri-drain 410	9	Well Point System (shallow) (6-9 m) Deep Dewatering Wells	SPT	Centroid of Primary, Secondary, Tertiary, and Quaternary Drop points	6,300	15.00	SMs MLs
	2	Primary	54	30	30		.	30	30									
	3	Secondary	54	30	30		.	30	30									
	4	Tertiary	54	30	20		.	30	30									
	5	Entire Area	18	12-18	2-3													

* Free-fall energy input equivalency; ** Not performed/necessary; *** Free-fall attempts unsuccessful Single-line used; **** Not adjusted for inflation; ***** Includes wicks and dewatering

pressures during compaction, and they would improve the drainage rate and consolidation time of the finer materials following dynamic compaction.

1.4 Evaluation of Liquefaction Potential

Pre-compaction SPTs, on approximately 15-meter centers, were conducted to define local site characteristics and to provide construction control. The potential for liquefaction in all materials was determined using Seed's empirical procedures (Seed, et. al., 1983 and Seed, et. al., 1984). A corrected blow count, $(N1)_{60}$, of 20 was necessary to preclude liquefaction. Within the desired treatment depth, approximately 50-percent of the preconstruction and pre-compaction sampled intervals did not meet the $(N1)_{60} \geq 20$ criteria.

1.5 Assessment of Treatment Effectiveness Using SPTs

A comparison of the pre- and post-treatment SPT blow counts and liquefaction potential to assess remediation effectiveness was made on the basis of:

A. Comparison of blow count change in sample interval elevations where a match in soil properties was possible in proximate drill holes.

B. Statistical presentation of blow count changes as a function of location within sectors and as a function of material type.

C. Comparison of liquefaction potential in all sampled intervals for all drill holes to the desired depth of treatment.

As each heavy tamping dynamic compaction phase was completed, companion SPT's were located adjacent to pre-compaction control SPT's to determine the effect of the previous treatment. This process was used despite recognition that it may take some time to register the phase's full effect. Nevertheless, the tests were made to try to determine whether any positive benefit was being produced, and if the necessary compaction was being accomplished with only a portion of the total specified compaction effort.

Despite the fact that the post-compaction holes were drilled within two meters of the pre-compaction drill holes, the stratigraphy based on laboratory gradations performed on each SPT interval did not match perfectly. As illustrated on Figure 3, the arrows connect what are determined by judgment to be sample intervals with equivalent material properties in the companion holes. On the basis of this system of comparison, approximately 500 matches were made between pre-and-post-compaction drill holes following treatment.

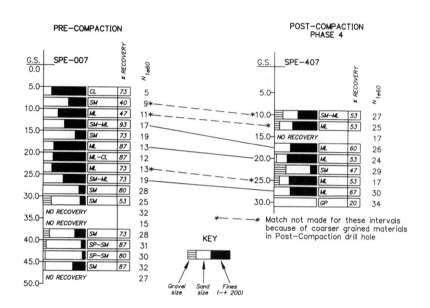

Figure 3. - Typical Pre-to-post-compaction Comparisons

Post-treatment SPTs when compared to the required $(N1)_{60}$ criteria of 20, indicated only isolated intervals remained potentially liquefiable. The exception was a relatively continuous silt zone at a depth of about 6 to 8 meters in Sectors B through E of the Stage I treatment area. This zone had a fines content around 85 percent and post-compaction $(N1)_{60}$ values averaging 16 to 18. Although they did not meet the desired blow count, the material was judged to be non-liquefiable based on fines content. This assumption is considered valid based on examination of Seed's empirical curves (Seed, et. al. 1984) of required $(N1)_{60}$ blow count for materials with fines. A summary of the improvement in SPT blow count by material type for sectors B-H (Stage I) following phase 4 treatment is given in Table 2.

TABLE 2. - IMPROVEMENT IN SPT BLOWCOUNT
BY MATERIAL TYPE AND SECTOR

SECTOR	GP and SP	SM	ML	Other	All
B	12.5(10)	7.2(5)	4.4(18)		7.7(33)
C	12.3(15)	9.4(5)	7.4(11)	11.8(4)(SM-ML)	10.3(35)
D	10.7(10)	6.2(5)	5.3(9)		7.7(24)
E	15.6(8)	11.3(21)	5.5(17)		9.9(46)
F	11.1(10)	14.5(6)	7.3(10)	7.7(2)(SM-ML)	10.2(28)
G	14(8)	10.0(3)	7.0(2)		12.0(13)
H	19(19)	10.5(8)	6.3(5)		15.5(30)
Summary	14.0(80)	10.4(53)	5.8(70)	10.4(6)	10.3(209)

(N) = Number of sampled intervals

The following conclusions were reached based on SPT results following foundation treatment:

A. A significant increase in SPT blow count occurred as the result of dynamic compaction.

B. In general, each additional phase of compaction resulted in additional improvement.

C. The depth of recognizable or apparent effect of treatment varied from 6 to 15 meters. Most observations indicated remediation for liquefaction to at least 9 meters.

D. Improvement was not uniform from interval to interval. In fact, the estimated improvement showed considerable irregularity.

E. High fines content materials were difficult to treat. However, with appropriate fines content adjustments to established $(N1)_{60}$ criteria, acceptable levels of improvement were realized.

1.6 Evaluation of Wick Drains

The similarity of stratigraphy in sectors B through E and the presence of a thick fine-grained layer at shallow depth provided an excellent opportunity for a large scale field test of the wick drain effectiveness. Areas B and C, which would have a low embankment height and little water stored behind the embankment, were not provided with wick drains, while the fine-grained portions of areas D through H were provided with wick drains.

Evaluation of the wick drain effectiveness was made on

both a quantitative and qualitative basis. Comparison of the improvement of just the matched materials in sectors B and E showed that gravels improved by 25 percent, silty sands improved by 57 percent, and silts improved by 25 percent more in sector E than did these materials in sector B. The presence of the wick drains doubled the blow count improvement in the fine-grained zone from an increase of 4 blows to an increase of 8 blows.

The silty-sand and gravel zones immediately below the fine-grained zone showed no improvement in sector B, where no wick drains were used. In sector E, where wick drains were used, the average blow count increased below the fine-grained zone from 22 to 28 blows. Therefore, wick drain use was found to benefit the dynamic compaction program at Jackson Lake. However, the cost and benefit of requiring wick drains in future projects still need to be viewed in relation to the amount of necessary treatment.

1.7 Effects of Time (Aging) and Plasticity on Improvement

Stage I dynamic compaction was completed in late autumn of 1986. Eighteen SPT's were performed to assess time (aging/consolidation) effects during the spring of 1987. Approximately four to six months of consolidation time in a cold climate occurred between the two sets of results.

In areas where comparisons could be made, large improvements were noted at eight locations, small improvements were noted at three locations, and inconclusive or no improvement was observed at five locations. Overall, 34 $(N1)_{60}$ values in the fine-grained zone were below 20 immediately after compaction and 10 $(N1)_{60}$ values were below 20 after the resting period.

The silt zone identified in sectors B through E was also examined to determine the effect plasticity may have had on the densification success from dynamic compaction. Data collected clearly indicated: a higher percentage of the more plastic zones failed the $(N1)_{60}$ blow count criteria, and additional time upon completion of dynamic compaction resulted in a distinct improvement of non-plastic soils and had less of an impact on the more plastic soils.

2.0 MORMON ISLAND AUXILIARY DAM MODIFICATIONS

2.1 Project Description

Mormon Island Auxiliary Dam (Mormon Island) was designed and constructed by the U.S. Army Corps of

Engineers (COE) in the late 1940's and early 1950's. Upon completion, ownership was transferred to Reclamation for operation and maintenance. The structure is a zoned earthfill embankment dam 50 meters high at the maximum section and 9 meters wide at the crest (see Figure 4).

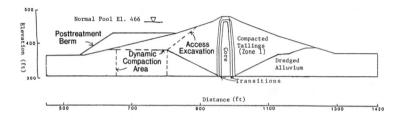

Figure 4. - Mormon Island Auxiliary Dam and Phase I Modification Features

Total length of the embankment is approximately 1,470 meters. The narrow, central impervious core is a well compacted clayey mixture and is founded on bedrock over the entire length of the dam. The core is flanked by well compacted transition zones on both the upstream and downstream sides. The inner transition is composed of silty sand from decomposed granite. The outer transition was constructed with the minus 5 centimeter fraction of the dredged alluvium.

Mormon Island was constructed across the Blue Ravine, an ancient channel of the American River. At the dam site, Blue Ravine is more than 1.6 kilometers wide and is filled with about 20 meters of alluvial deposits. Prior to construction of Mormon Island, the deepest portion of Blue Ravine was repeatedly dredged for gold. The dredging process tended to deposit the tailings in a very loose condition, with the finer materials concentrated near the bottom of the dredged channel and the coarser materials near the surface. Sloughing of unstable tailing piles at the dredge pond surface resulted in interlayering of the fine and course materials.

Part of Mormon Island's embankment shells (GPs with cobbles), approximately 275 meters in length, is founded on the loose dredged alluvium. Due to the coarse nature of these materials, the Becker Hammer Penetration Test (BPT), as defined by Harder and Seed, was selected as the primary tool for estimating liquefaction potential based on penetration resistance. COE investigations (Hynes-Griffin, et. al. 1988) concluded that extensive liquefaction was expected and that catastrophic loss of reservoir may result under earthquake loading. Remedial measures were determined necessary for the portion of the dam where the embankment shells are founded on the dredged alluvium.

Drought conditions and low reservoir levels in 1990 provided an opportunity to perform remedial treatment for liquefaction upstream and under the embankment. Densification by dynamic compaction was selected based on time available for design and construction, technical feasibility, and overall cost. Studies (Ledbetter, et.al., 1991) defined treatment requirements to 18 to 20 meters. To address the probability of not being able to treat below 12 meters by dynamic compaction, a berm was added above the dynamic compaction treatment area to allow access so that future remediation below 12 meters can take place without lowering the reservoir. A summary of remediation features is provided in Figure 4.

2.2 Compaction Program

In order to attempt treatment at depths of 15 meters or greater, primary drop points were spaced at 15 meters and received more than 30, 33 meter high tamper impacts. This compares to a spacing of 12 meters for at Jackson Lake Dam (refer to Table 1). Secondary drop points (30 impacts) split the primary spacing and tertiary drop points split secondaries (15 impacts). A final "ironing" phase, with two drops from 9 meters on a continuous side-by-side pattern was also performed to help densify the disturbed surface. The initial ironing phase was determined unnecessary.

Based on previous experience and knowledge of potentially available equipment, the specifications required an impact energy per drop equivalent to a 35 ton tamper "free-falling" 30 meters. The contractor used Dynamic Compactors and two tampers to accomplish the work. The tampers, weighing about 35 tons, were circular in shape, roughly 2 meters in diameter at the base, and were made of steel.

To confirm energy input requirements, the contractor performed tests with a laser system to measure velocity of the tamper just prior to impact at the ground surface. A drop height of around 33 meters was determined necessary to achieve the 30-meter free fall energy input requirement.

The construction procedure used was to drop on any given impact point until the resulting hole, or "crater", was about 2.4 to 3.7 meters deep. Initial crater diameters were around 3.7 to 4.6 meters. This could take as few as 6 drops early on in the program and as many as 15 drops with increase in density.

Specifications precluded dropping tampers when water levels in the area were less than 1 meter below the crater depth elevation. Craters required backfilling to meet this criteria and allow efficient removal of the tamper from craters prior to completion of all specified drops at all of the primary and secondary impact points. Some tertiary impact points could receive the full 15 drops specified without requiring crater backfilling.

Backfill consisted of highly pervious dredged alluvium materials from the embankment and foundation excavations. The backfilling operation took about 5 to 15 minutes. Compaction was completed at each impact location prior to moving to another impact location. At Jackson Lake and Steinaker Dams, all specified drops were not required to be completed at any impact location prior to moving on to another impact location. Cycle times between drops ranged from about 80 to 90 seconds.

2.3 Verification Testing

Thirty BPTs, seven SPT's, and one crosshole shear wave velocity triplet were completed from the compaction working surface to bedrock prior to performing dynamic compaction. Seventeen BPTs were performed after completing dynamic compaction at primary impact point locations. Forty-two BPTs and five companion SPTs were performed within two weeks following compaction construction. These post-construction BPTs were performed at evenly distributed (± 15-meter spacing) locations across the site at primary, secondary, tertiary and quaternary impact locations.

Three to four months after compaction, seven BPTs and three companion SPTs were performed through the berm in an attempt to quantify the aging effects on treatment and obtain additional BPT to SPT comparison data. Note that the berm added about 17 meters to explorations performed previously. Downhole geophysical testing, consisting of natural gamma, neutron, gamma-gamma (density), and induction (electrical conductivity) testing, was also completed during this time period.

About one year after dynamic compaction, four BPTs along with companion SPTs were completed through the berm. This work was done to determine the effect of friction on the Becker Hammer casing and BPT results. Two af the BPT locations were cased through the berm to the top of the remediation zone and two were cased to about a depth of 9 meters within the remediation zone. Post-compaction crosshole shear wave velocity testing was also completed at 3 locations within the remediated zone. The crosshole sites were reoccupied and tested in the spring of 1994 to assess lateral stress redistribution, time/aging and consolidation effects.

2.4 Penetration Testing Results

Generally, a very high level of compaction effectiveness was observed in the upper 9 to 12 meters of the treatment zone. Below 12 meters less dramatic, but still significant, increases in penetration resistance were noted (Stevens, et. al., 1993). The middle one third, or 15-meter width of the 45-meter wide treatment area, indicated higher degrees of treatment below 12 meters than the outer two thirds of the area.

Preliminary evaluation of BPT testing results (Stevens, et. al., 1992) raised questions on the validity of BPT data reduction procedures on data collected through the very dense remediated dredged alluvium and compacted berm materials due to the potential effects of friction on the test. Subsequent to these investigations, a comprehensive assessment of all BPT, SPT and crosshole shear wave velocity data was performed on data collected at the site (Harder, 1993). The assessment showed that:

   A. SPT data in sandy materials allowed a site specific calibration for the BPT.

   B. Different rig factors needed to be applied to each rig to account for drill rig efficiencies different from those interpolated using "standard" procedures.

   C. Post-compaction SPT and BPT data indicate that the increase in BPT blow counts observed in deeper portions of the treated area <u>were the result of the densification process</u> and not from apparent friction effects.

   D. Crosshole shear wave velocity data show moderate to low increases in densification to depths as deep as 16 meters.

   E. Significant casing friction effects were evident in BPT data collected through the highly compacted upstream berm and need to be considered in data reduction.

   F. After incorporating the above factors the BPT data reduction procedures recommended by Harder and Seed are considered applicable for this site.

2.5 Effect of Location on Penetration Testing Results

BPT testing locations were uniformly distributed across the 45-by-275 meter treatment area near nine primary, eight secondary, ten tertiary, and fourteen quaternary impact locations. This amount of testing, the averaging of results, and the relative uniform nature of the dredged alluvium materials across the site revealed that a study of the effects of location on penetration testing can yield tenable results. Presented on Figure 5 is a summary of average BPT values for all pre-compaction and post-compaction-by-location BPT testing.

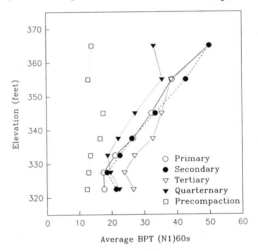

Figure 5. - Average BPT $(N1)_{60}$s at Various Testing Locations

Review of this information suggests the following for this site and the compaction program used:

    A. Penetration testing results near tertiary impact locations may yield higher than average estimates of ground improvement by dynamic compaction.

    B. Overestimating effectiveness of ground improvement in the upper reaches of the compaction area may result from testing near primary and/or secondary impaction locations.

    C. Testing near primary and/or secondary impact locations may result in underestimating ground improvement in the lower reaches of the compaction area due to rafting effects.

    D. Testing results near quaternary impact locations may indicate lower than average ground improvements in upper portions of the remediation area, but should appropriately define average treatment effectiveness at depths greater than 12 meters.

E. Testing at or near quaternary impact locations to verify ground improvement by dynamic compaction is appropriately conservative for high-risk facilities such as dams with liquefaction-related stability concerns.

2.6 Shear Wave Velocity Testing Results

As depicted on Figure 6, a considerable increase in shear wave velocity was realized by the dynamic compaction program. At the only pre-compaction to post-compaction comparison location, the average increase in shear wave velocity was about 280 m/s above a depth of 12 meters and 52 m/s from about 12 to 17 meters. By comparison, average shear wave velocity increases measured at three sites at the Jackson Lake project were on the order of 52 m/s in the upper 12 meters and about 15 m/s from 12 to 15 meters (from Sirles, et. al., 1990). The difference in improvement can be explained in part as follows:

A. The total amount of energy input per area of treated ground was much larger for the Mormon Island project than it was for the Jackson Lake project. The design objective at Mormon Island was to treat to as great a depth possible. A 9- to 12-meter effective treatment depth was the design intent for the Jackson Lake Dam modifications.

B. The materials receiving treatment at Jackson Lake were generally finer grained than those at Mormon Island. Also, the percentage of free-draining materials available to assist with drainage from finer materials during treatment was greater at the Mormon Island project.

C. Post-construction shear wave velocity data, as presented on Figure 6, were taken through a 15 to 17-meter very dense working platform berm for the Mormon Island project. No adjustments were made to the data to compensate for increased overburden effects. The Jackson Lake pre-and post-construction data were taken under similar effective stress conditions.

To evaluate lateral stress redistribution and time dependent treatment and overburden consolidation effects for a final assessment of the structure, the three post-construction crosshole triplet installations at Mormon Island were reoccupied and tested in the spring of 1994. This was two years after the initial crosshole testing and after a complete reservoir refilling cycle had taken place.

As presented on Figure 6, considerable increase in shear wave velocity took place between 1992 and 1994.

Average shear wave velocity increases on the order of 45 m/s were regularly observed across the site. An important general observation of the data indicates that finer grained materials increased less with time/consolidation effects than the coarser materials. The reservoir water surface at the time of the 1992 testing was about 12 meters higher than for the 1994 testing. This would suggest that the values obtained in 1994 are conservative in assessing increases in shear wave velocity since no adjustment for the higher water surface was made. SPTs were in progress at the time of this writing and thus could not be compared to the shear wave velocity information.

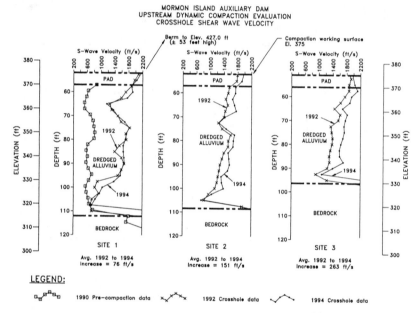

Figure 6. - Crosshole Shear Wave Velocity Testing Results

3.0 STEINAKER DAM MODIFICATIONS

3.1 Project Description

Steinaker Dam was built as a zoned, earthfill dam in 1960 on a stream cut through the flank of an anticline. The cap rock forming the hogback ridge of the abutments is an erosion-resistant sandstone while the majority of the abutment bedrock foundation is shale. Alluvial materials form the foundation beneath the main dam section.

The alluvial sequence consists of a 6-meter thick lean clay, underlain by 11 meters of interbedded fine, clean sands and non-plastic silts. A stiff, fat clay is found at the bottom of the sequence in the deepest valley section, sometimes lying directly on bedrock, and sometimes underlain by a poorly-graded sand. Pump-out tests indicated separate aquifers above and below the fat clay layer. Groundwater in the upper aquifer fluctuated around an elevation near the bottom of the surface clay layer. Groundwater below the fat clay was artesian, rising to a level slightly above the original ground surface.

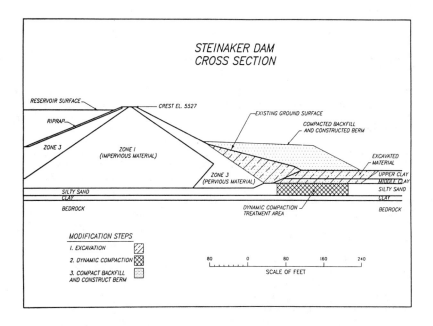

Figure 7. - Typical Section - Steinaker Dam Modifications

The interbedded silts and sands were recognized as being "quick" during the original dam construction. The original dam design called for a cutoff trench to be constructed to bedrock beneath the core. However, artesian groundwater was encountered, and it was decided to remove only the low-strength surface clay layer beneath the dam's core. Once the excavation encountered the sands, SPT's were conducted to determine if any clay layers remained. With none being encountered, the foundation was accepted. It was thought that the silts and sands would settle under the weight of the dam.

Testing conducted in 1986 under the safety of dams program determined that the silts and sands beneath the dam are liquefiable, and that the dam would fail should the maximum credible earthquake (Magnitude 6.5) occur. Alternatives were investigated and the remedial modification selected was a stability berm placed on treated foundation material. The modification entailed excavating 8 meters of material at the downstream toe, treating the remaining 9 meters of liquefiable sands and silts using dynamic compaction, and constructing a berm above the treated material, as depicted by Figure 7.

3.2 Construction

The principal work components performed for the modification of Steinaker Dam included the following: Deep wells were installed and pumped to dewater the excavation downstream from the existing dam's toe. Approximately 9 meters of clay and sand foundation materials were removed. A 1.5-meter thick compaction pad was constructed using 15-centimeter minus, clean material. Perimeter wellpoints were installed to lower the water table at least 3.7 meters below the top of the compaction pad and to assist in providing drainage during dynamic compaction. Wick drain were installed on 1.5-meter centers to a depth of 9 meters. The wick drains terminated at 9 meters to prevent communication between the two groundwater aquifers. Dynamic compaction commenced using an energy equivalent of dropping a 30 ton weight from 30 meters. A stability berm was constructed over the densified material.

3.3 Compaction Program

The design parameters for dynamic compaction relied heavily on experience gained from the Jackson Lake and Mormon Island projects. The grid spacing, drop weight and height, and the number of drops per drop location were all chosen knowing the same parameters successfully achieved 9 meters of treatment at both locations. Unknown were the need for an ironing phase, the amount of material needed for crater backfill, and the degree of improvement that would be obtained in the plus-35-percent fines silty-sands and sandy-silts.

The initial ironing phase was included in the modification specifications for Steinaker Dam even though it was were not needed at either Mormon Island or Jackson Lake. The specifications called for 2 or 3 drops from 12 to 18 meters at each primary, secondary, and tertiary location. The need for the ironing phase and the drop height and number were determined during energy delivery testing.

Energy delivery testing was required because the contractor chose to keep the weight tethered during drop cycles. Reclamation specified a 30-ton weight to be dropped in free-fall, from a height 30 meters above the surface of the compaction pad. However, with the weight tethered, the contractor was allowed to drop from a height greater than 30 meters provided he could demonstrate equivalent energy delivery upon impact.

The energy testing took place in the center of the compaction area. During the first drop, the weight sank approximately 4 meters, dramatically illustrating the need for the ironing phase.

The dynamic compaction was conducted in a 45 by 135 meter grid. Primary and secondary drop points each received 30 drops while tertiary drops received 20 drops. Primary and secondary drop points were located at the corners of squares, 7.6 meters to a side. Tertiary points were located in the center of each square. In each row or column of the grid, primary and secondary points were staggered at 7.6 meter spacing.

The drop sequence was tightly controlled at Steinaker Dam. All 30 drops at primary locations were completed before drops were allowed at secondary locations. The same was true of the secondary and tertiary phases. The contractor was directed to work across one complete row and back the next instead of conducting all drops in one area. This was done to inhibit pore pressure buildup in any one area.

Within each phase, Reclamation intended to have the required number of drops completed in two or three passes. The number of drops per pass depended on crater development. The wellpoints brought the water table down at least 3.7 meters below the compaction pad surface so dropping the tamper into craters deeper than 3.7 meters would not be effective. At the beginning of the primary phase, the contractor was directed to drop the tamper until the crater formed to a depth of about 3.7 meters. After a few drop point locations were observed, 15 drops per pass was found to work well.

Before the contract was awarded, Reclamation estimated 17,600 cubic meters of material, or approximately 30 percent of the total volume requiring treatment, would be needed to backfill craters. The actual amount used was 12,400 cubic meters, or approximately 21 percent of the total treatment volume. The breakdown per phase is given in Table 3. No ground heave was observed at any time during the treatment at Steinaker Dam. Wet areas were observed where wick drains exited the ground surface.

Table 3. Breakdown of Crater Backfill Volumes by Phase

Preliminary Ironing	(2)	1,894 m^3
Primary Drops	(30)	2,912 m^3
Secondary Drops	(30)	4,422 m^3
Tertiary Drops	(20)	2,610 m^3
Final Ironing	(2-3)	563 m^3

3.4 Verification Testing

SPTs were used as the sole means of treatment verification. The tests were all conducted at locations midway between primary and secondary drop locations. It was thought that these locations would experience the least amount of compaction effort. Pre- and post-compaction drill holes were located approximately 1.5 meters from each other. The 45 by 135 meter treatment area was divided into twelve squares with 23 meters to a side. Two SPT drill holes were placed in each square before the treatment. Somewhat less than two drill holes per square were obtained after treatment because of construction scheduling problems.

The SPTs for compaction monitoring were concentrated between elevations 1633 and 1629, although pre-compaction SPTs extended below 1629 to improve understanding of the locations of the fat clay and the bedrock. Elevation 1633, being 4.5 meters below the compaction pad surface, was chosen because it was thought that material above this elevation would be thoroughly compacted and would contain compaction pad materials. Descriptions of drilling conditions from driller's logs confirm that the gravel and cobble compaction pad materials ceased in the vicinity of elevation 1633.

Pre-compaction SPTS were conducted from an excavated surface at elevation 1639.5. A 3,000-year-old village site was discovered in test pits during design-phase explorations. This required a two-month halt in construction for archeological investigations. Pre-compaction SPT's were conducted during these archeological investigations.

Post-compaction SPTs were conducted from ground surface elevations 1638 and 1645. The original plan for post-compaction verification testing called for 24 drill holes. SPT's were to be conducted in a 4.5 meter interval starting 4.5 meters below the surface. Government drill crews were given fourteen days access to the compaction pad to conduct the SPT's. Only fourteen drill holes were completed from this elevation during this time period.

Preliminary indications from the fourteen holes were that the treatment had achieved the desired compaction. However, three squares did not have any post-compaction verification testing. The excavation was filled back to the original ground surface before the construction was shut down for the winter. A decision was made to perform the additional SPT's after the excavation was refilled.

SPTs conducted from elevations 1638 and 1645 would encounter significantly different stress states. A test was conducted to see if applying the blow count correction factor (C_n) would produce comparable post-compaction blow count values. Two SPT holes were drilled and tested from elevation 1645 in the vicinity of drill holes previously tested from elevation 1638. The comparison showed that corrected post-compaction blow count values were not significantly different when tested under different overburden stress states.

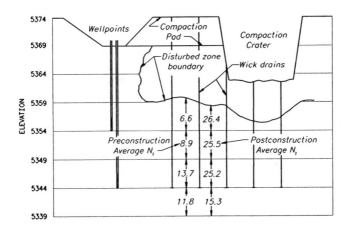

STEINAKER DAM SPT VERIFICATION RESULTS

Figure 8. - Dynamic Compaction Features with Pre-to-Post-compaction SPT Results

The dynamic compaction achieved the desired densification throughout the treatment area. Figure 8 shows the SPT results for pre- and post-compaction. The foundation was divided into 1.5-meter thick layers and the average corrected blow count is shown for each layer. Uniform treatment was achieved throughout the area to a depth of 9 meters from the compaction pad surface. There were no SPT intervals between depths of 4.5 to 9 meters that remained potentially liquefiable after treatment.

Below a depth of 9 meters three materials were encountered. Approximately 25 percent of the area was bedrock. Another one third of the area was comprised of the fat clay which is not liquefiable. Liquefiable and non-liquefiable silty sands comprised the remaining 42 percent of the area tested using SPT's. All the liquefiable intervals from this depth were found in drill holes located in two isolated areas of the foundation.

The silty sands above 9 meters went from an average of from 6 to 10 blows to an average above 25 blows, the average blow count in silty sands below 9 meters went from 11.8 to 15.3. Three hypotheses for this behavior are: 1) nine meters of treatment is all that could be expected from the drop specifications used, 2) the fat clay somehow reflected or absorbed energy, or 3) the wick drains (which were only installed to the 9-meter depth) were extremely effective. Data indicate the latter reason.

The increase in average blow count for each 1.5-meter zone did not show a regular decrease with increasing depth. If there was a decreasing energy delivery with depth it should be accompanied by a decreasing effectiveness demonstrated by SPT results. Then, the 9 meters of treatment might have been all that could be expected with the drop specifications used. Instead, a sharp boundary was observed between the three zones where blow counts increased to approximately 25 blows and the zone where the increase was only to 15 blows.

Some areas bounded by the underlying fat clay did undergo a blow count increase to the mid twenties. Therefore, the absorbed energy hypothesis is probably not correct.

The wick drains were only installed to a depth of 9 meters. Wick drains were specified on 1.5-meter centers at Steinaker because the materials requiring treatment had an average fines content of 45 percent. Experience at Jackson Lake Dam showed that the non-plastic silty materials would not see significant improvement without the wick drains. The fact that the drains stopped at the same depth where the degree of improvement significantly decreased is another demonstration of their effectiveness in these fine-grained materials.

4.0 CONCLUSIONS

Deep dynamic compaction can remediate liquefiable soils to acceptable levels and depths at both fine- and coarse-grained sites, provided appropriate procedural and pore pressure dissipation features are used.

Providing pore water pressure relief measures can enhance the dynamic compaction by any of the following means:

A. Install wick drains to the full depth requiring treatment.

B. Construct drainage trenches around as much of the area as possible for surficial and perimeter pore pressure dissipation, water control and as an exploratory and construction observation tool.

C. Specify drop sequencing and pattern on a row-by-row basis.

D. Increase the number of phases of the compaction program to require more equipment movement. This provides time for pore pressure dissipation to take place and for materials to consolidate.

E. Reduce energy-per-drop requirements in shallow or confined areas.

The inner portions of the treatment area will receive a higher degree of treatment than outer portions at depths greater than about 9 meters. Complete liquefaction remediation below a depth of 12 meters may be questionable, depending on the level of improvement required at depth. Increases in penetration resistance below a 12 meters depth can be achieved with specialized equipment and implementation of appropriate procedures (ie., by adding drops to the compaction program at quaternary impact locations, or by adding drops at existing impact locations). The amount of increased penetration resistance is unknown and probably highly site specific.

Consider reducing or eliminating drops during later phases of compaction if pore pressure instruments and/or field observations indicate induced pore pressures are restricting the compaction process. Penetration testing subsequent to each phase of the program may be beneficial depending on contract and schedule flexibility.

Increases in the shear wave velocity and penetration resistance can be expected with time. Verification testing performed shortly after completion of compaction will be conservative in estimating total effectiveness of treatment. Redistribution of lateral stresses with time would not be expected to occur to the extent of effects associated with aging or consolidation. Finer-grained materials will increase less with time than will coarser-grained materials. Materials exhibiting plasticity will increase less with time than non-plastic materials.

Wick drains can improve treatment effectiveness, though the benefits appear to be dependent on site characteristics and on wick drain materials. Only small increases in treatment effectiveness shown for the Jackson Lake project. Considerable and undeniable benefits were observed at the Steinaker project. Thus, installation of wick drains should be considered in specifying projects.

ACKNOWLEDGEMENTS

The authors would like to acknowledge the many individuals within the Bureau of Reclamation (Denver, Mid-Pacific, Pacific Northwest, Salt Lake City, Bend, Provo, Central Snake and Folsom Project Offices) who participated in the successful completion of the above projects. Also, all the Corps of Engineer's personnel from the Sacramento District, San Francisco and Waterways Experiment Station Offices are greatly appreciated for their participation and input to design and construction of the Mormon Island Auxiliary Dam modifications.

REFERENCES

Farrar, J.A., Wirkus, K.E. and McClain, J., (1990), "Foundation Treatment for the Jackson Lake Dam Modification," Proceedings of the U.S. Committee on Large Dams, New Orleans, LA., March 14.

Farrar, J.A., Yasuda, N., (1991), "Evaluation of Penetration Resistance for Liquefaction Assessment at Jackson Lake Dam", Proceedings of the 23rd Joint Meetings, Wind and Seimic Effects, NIST SP 820, 510 through 526.

Harder, L.F., (1993), "Evaluation of Becker Hammer Soundings Performed at Mormon Island Auxiliary Dam in Conjunction with Upstream Remediation", Prepared for the U.S. Army Corps of Engineers, Sacramento District, Sacramento, California.

Hynes-Griffin, M.E., (1987), "Seismic Stability Evaluation of Folsom Dam and Reservoir Project, Report 1: Summary Report," Technical Report GL-87-14, U.S. Army Engineer, Waterways Experiment Station, Vicksburg, Mississippi.

Hynes-Griffin, M.E., Wahl, R.E., Donaghe, R.T. and Tsuchida, T., (1988), "Seismic Stability Evaluation of Folsom Dam and Reservoir Project, Report 4: Mormon Island Auxiliary Dam - Phase I," Technical Report GL-87-14, U.S. Army Engineer, Waterways Experiment Station, Vicksburg, Mississippi.

Luebke, T.L., McDaniel, T., Stevens, M.G., 1985, Official Correspondence, Travel Report, "Site Visit to Steel Creek Dam", Discussions with U.S. Army Corps of Engineers and Geotechnical Engineers, Inc., on dynamic compaction procedures implemented at Steel Creek Dam.

Ledbetter, R.H., Finn, W.D. Liam, Nickell, J.S., Wahl, R.E. and Hynes, M.E., 1991, "Liquefaction Induced Behavior and Remediation for Mormon Island Auxiliary Dam," Proceedings from the International Workshop on Remedial Treatment for Liquefiable Soils, U.S. and Japan Panel on Wind and Seismic Effects, Tsukuba, Science City, Japan.

Seed, H.B., Idriss, I.M., and Arango, I., 1983, "Evaluation of Liquefaction Potential Using Field Performance Data," Journal of Geotechnical Engineering, ASCE 109(3):458-482.

Seed, H.B., Tokimatsu, K., Harder, L.F., and Chung, R.M., 1984, "The Influence of SPT Procedures in Soil Liquefaction Resistance Evaluations," Report No. UBC/EERC-84/15, Earthquake Engineering Research Center, University of California, Berkeley, California.

Sirles, P.C., Viksne, A., (1990), "Site Specific Shear Wave Velocity Determination for Geotechnical Engineering Applications", Society of Exploration Geophysicists Specialty Publication on Environmental and Engineering Geophysics, Vol. III.

Stevens, M.G., Farrar, J.A, Allen, M.G. and Von Thun, J.L., (1992) "In Situ Testing Performed at Jackson Lake Dam and Mormon Island Auxiliary Dam," Second Interagency Symposium on Stabilization of Soils and Other Materials, New Orleans, Louisiana.

Stevens, M.G., Allen, M.G. and Farrar, J.A. (1993), "Construction and Verification of Ground Improvements at Mormon Island Auxiliary Dam," ASCE, Geotechnical Special Publication No. 53, Geotechnical Practice in Dam Rehabilitation, 961 through 968.

U.S. EXPERIENCE WITH DYNAMIC COMPACTION OF COLLAPSIBLE SOILS

Kyle M. Rollins[1] and Ji-Hyoung Kim[2]

Abstract

During the past 10 years dynamic compaction has been used to improve collapsible soils on nine projects in five western states. In this paper, the soil properties, compaction procedures and subsequent performance at each site are summarized and compared. Specifications used in conducting the work and methods used for evaluating improvement are also outlined. Finally, correlations are presented for estimating the depth of improvement, the depth of craters, and the level of vibrations based on measurements made at the various sites.

Introduction

As development continues throughout the western United States, buildings and highways must be constructed over collapsible soils. Collapsible soils are stiff and strong in their dry natural state, but lose strength and undergo significant settlement when they become wet. Settlements associated with collapsible soils can lead to expensive repairs if the soils are not treated in some way prior to construction. During the past 10 years, dynamic compaction has been employed to improve the strength and decrease the settlement potential of collapsible soils on nine projects located in five western states as shown in Figure 1.

The objective of this paper is to summarize information from available case histories involving deep dynamic compaction (DDC) of collapsible soils in the United States and present empirical correlations based on the data. While a number of correlations have already been developed for soils in general, some refinements may be necessary for collapsible soils. Collapsible soils exist in a very loose meta-stable

---

[1] Assoc. Professor, Civil Engineering Dept., Brigham Young Univ., Provo, UT 86402

[2] Engineer, American Engineering Co., 21442 North 20th Ave, Phoenix, AZ 85027

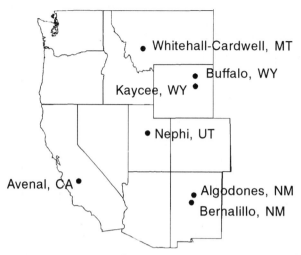

Figure 1 Locations of case histories involving dynamic compaction of collapsible soils in the western United States.

state, yet they are relatively incompressible in their dry natural state. These peculiar characteristics may lead to variations from conventional soil behavior during dynamic compaction. If existing correlations can be validated or more accurate correlations can be developed for collapsible soils, engineers will be in a better position to write specifications for DDC projects.

Description of Test Sites

The case histories discussed in this paper are: (1) A test section and full-scale project on I-90 between Whitehall and Cardwell, Montana in 1984, (2) a test section and full-scale project on I-25 near Algodones, New Mexico in 1985, (3) a test section and full-scale project at a state prison near Avenal, California in 1985, (4) two test sections and three full-scale projects on I-25 between Kaycee and Buffalo, Wyoming in 1989-1990, (5) a full-scale project on I-25 near Bernalillo, New Mexico in 1992, and (6) two research test projects in Nephi, Utah in 1990 and 1992. A summary of each project is given below which describes the site conditions, compaction parameters, and methods which were used to evaluate the degree of improvement achieved. Additional information on each site is summarized in Table 1.

I-90 between Cardwell and Whitehall, Montana

Interstate Highway 90 in southwestern Montana contained bumps, cracks, and pavement distortion, with the area between the Whitehall Interchange and the Cardwell Interchange being the most severely distorted. Field investigations

Table 1 Summary of soil conditions and dynamic compaction parameters on six projects involving collapsible soils.

Project	Whitehall-Cardwell, MT	Algodones, NM	Avenal, CA (test area)	Buffalo-Kaycee, WY	Bernalillo, NM	Nephi, UT
Treated area (m²)	93,100	15,900	800	119,560	51,340	107
Collapsible soil thickness (m)	6.0-7.6	6.0-7.6	6.0	7.6	6.0-7.6	6.0-7.6
Energy per drop (t-m)	291	374	226-282	415	648	89-111
Primary drop spacing/tamper diam.	2	1.77	1.36	1.77-2.13	2	1.77-1.88
Improvement depth (m)	7.9	7.3	4.0	7.6	6.0-7.6	4.1
Compactive energy per surface area (t-m/m²)	174	271	238-384	262-298	426	131-209
Compactive energy per volume (t-m/m³)	21.9	37.2	58.7-94.9	34.2-39.0	55.7-69.4	31.8-50.9
Contact pressure (kPa)	98	55	31	78	67	24-34
Cost per treated area treated ($/m²)	8.36	6.00	Not available	5.4-7.47	9.32	Not applicable
Cost per volume ($/m³)	1.04	0.82	Not available	0.65-0.98	1.53	Not applicable
Specification type	Method	Method	Performance	Method	Performance	Method

indicated that the pavement distress was a result of collapsible soils within the subgrade and that dynamic compaction would be the most economical method for treating the soils.

The project area is located in an intermountain basin and the collapsible soils of concern are located within Quaternary alluvial fan deposits laid down by intermittent stream flow as it enters the basin (Yarger, 1986). The collapsible fan deposits are typically silty sands or sandy silts with small amounts of clay and gravel. Borings showed that the collapsible soils beneath the distorted pavement generally extended to a depth of 6 to 7.6 m (20 to 25 ft). The dry unit weights were generally between 14.1 and 15.7 kN/m^3 (90-100 lb/ft^3) and natural water contents were around 10 percent.

Based on preliminary work on a test section, a compaction method was specified requiring a 13.6 t (15 tons) weight dropped from a height of 21.3 m (70 ft). Prior to dynamic compaction, the asphalt surface course was removed but the gravel base course was left in place. Compaction was performed in two passes. For the primary pass, 5 drops were made at each corner of a 3.65 m (12 ft) square grid. The secondary pass consisted of 3 drops at the center of each grid. The specifications called for extra drops if the weight penetration on the final drop exceeded 10 percent of the total crater depth up to that point. Upon completion of the compaction the upper 0.6 m (2 ft) of loose soil was compacted by conventional procedures. A total area of 93,113 m^2 (111,300 yd^2) were treated in this manner at a cost of approximately $8.36 per m^2 ($7/yd^2).

After the dynamic compaction, the Standard Penetration Test was conducted to inspect the depth of improvement. The average increase in the SPT N values for the treated soils was between 5 and 12 blows/0.3 m. The average depth of improvement was about 7.9 m (26 ft). Dynamic compaction was effective on all but approximately 5 percent of the areas compacted. The designed depth of improvement was 6.1 to 7.5 m (20 to 25 ft), however, about 5 percent of the treated areas contained pockets of collapsible soils which extended 9.1 or 12.2 m (30 or 40 ft) below the ground surface. Poor pavement performance appears to correlate with the zones where these deeper collapsible layers were not improved by the dynamic compaction.

I-25 near Algodones, New Mexico

In the late 1970's, a 5.63 km (3.5 mi) long section of pavement on Interstate Highway 25 near Algodones, New Mexico became heavily damaged due to settlement of collapsible soils in the subgrade. Road maintenance crews attempted to keep the roadway level by applying additional asphalt overlays in sections where settlement had occurred. Borings indicated that the thickness of asphalt was over 0.28 m (11 in) in some areas. New Mexico Highway Department engineers estimated that the maintenance cost over a 5 year period for the section was $6/m^2 ($5/yd^2). A number of treatment methods including dynamic compaction were evaluated under field conditions in an effort to improve the strength and compressibility characteristics of the underlying soil. Based on the results of the test sections, dynamic compaction was selected as the most cost effective treatment and it was used for the full scale production run.

Interstate 25 runs through the Rio Grande valley which is a graben formed by rifting in the region. The Sandia Mountains are located immediately east of the valley and were produced by the faulting. The near surface geologic features along I-25 vary from alluvial apron or coalesced fan deposits along the major streams to terrace deposits. The alluvial fan deposits, which are often collapsible, accumulate along the borders of rivers and major streams as a result of intermittent stream flow from small arroyos (Lovelace et al., 1982). Because of the low annual precipitation in this region, the annual average depth of wetting is only between 0.6 to 1.0 m (2 to 3 ft). As a result, thick deposits of soil can form without ever being consolidated under saturated conditions.

The section of the roadway which showed the greatest settlement was located on apron deposits and the depth of collapsible soils was between 6 to 7.6 m (20 and 25 ft). The primary soil types in the profile were non-plastic silts and silty sands with minor strands of sandy gravel (Lovelace et al., 1982). Dry densities ranged from 11.3 to 14.3 kN/m^3 (72-91 lb/ft^3) and the natural moisture content varied from 5 to 7 percent. Based on oedometer collapse testing, the maximum settlement due to collapse was estimated to be about 0.6 m (2 ft), or about 8 to 10% of the collapsible soil thickness.

In 1982, a field test of dynamic compaction was made along with several alternative methods including vibro-flotation, deep mixing and compaction, and pre-wetting. The test section was located on a frontage road adjacent to the interstate in the middle of the apron deposits where the greatest settlement was detected. Dynamic compaction was found to be the most cost effective method for treating the collapsible soils. Dry densities were generally increased to between 14.9 to 16.5 kN/m^3 (95 to 105 lb/ft^3).

In 1984, full-scale production runs began on the interstate itself and dynamic compaction was used over 6 lane miles of roadway near the north end of the test section. The total treated area was 15,900 m^2 (19,000 yd^2) and the average cost was about \$6/m^2 (\$5/yd^2). Based on the test section, a tamper weight of 13.6 t (15 tons) was dropped from 20 m (65 ft). The compaction was done in one pass with 4.57-m (15-ft) print spacing. Every other row was offset from the previous row by 2.3 m (7.5 ft) and each drop location had 6 to 7 drops. The depth of improvement was found to be approximately 7.3 m (24 ft) based on SPT and CPT data.

While little post-compaction soil information is available, performance of the roadway section treated with dynamic compaction has generally been quite good

(Lovelace, Personal Communication, 1993). After about 10 years of service only a few sections, representing a small fraction of the treatment area, are showing abnormal settlement. These sections are generally associated with drainage locations where there is a greater likelihood of deep wetting. It is also possible that the thickness of collapsible soils extends beyond the depth of improvement in some of these sections.

State Prison near Avenal, California

In 1985, geotechnical investigations for a state prison near Avenal, California indicated that collapsible soils were present at the site. Dynamic compaction was performed to increase the soil density and stability prior to construction of the prison. The prison site is located on the Kettleman Plain, a northwest trending valley bounded by the Kreyenhagen Hills to the west and the Kettleman Hill to the east. The soils near the ground surface at the prison site were formed by alluvial fan deposits which slope gently downward to the east. The alluvial fan deposits primarily consist of fine to very fine silty sand, and fine sandy silt, with some layers of clay and gravelly sand (CH2M-Hill, 1985a).

Subsurface explorations involved drilling over 37 boreholes and excavating 49 test pits. Standard penetration tests indicated that the alluvial fan deposit extending to depth of about 6 m (20 ft) were in a loose to medium dense state with N value of 15 or less. Below a depth of about 6 m (20 ft), the sediments were medium dense to dense with N values typically greater than 30. Oedometer collapse tests were used to identify soils that were susceptible to changes in moisture content. Most samples tested showed settlement when saturated under load. The average strain upon wetting was about 6 percent and samples with initial unit weights less than about 14.1 kN/m^3 (90 lb/ft^3) settled more than 5 percent when saturated (CH2M-Hill, 1985a).

A test section was used to determine the feasibility of using dynamic compaction to densify the proposed site. The goal was to raise the SPT N values to 14 or greater which corresponds to a CPT tip resistance of approximately 5.88 MPa (50 tons/ft^2) or greater.

Two test sites were divided into four 9.1 m x 9.1 m (30 ft x 30 ft) quadrants and in each quadrant a different compactive energy level was used as shown in Table 2. The tamper weighed 12.4 t (13.6 tons), and its dimensions were 2 m x 2 m x 0.46 m (6.5 ft x 6.5 ft x 1.5 ft). Primary drops were spaced approximately 3 m (10 ft) on centers with secondary drops spaced between them. Subsurface characterization before and after the DDC consisted of cone penetrometer soundings and SPT borings at each test quadrant. The CPT and SPT holes were positioned 10 diameters away from each other to minimize interference. Significant improvement was observed to depths of about 4 m (13 ft) (CH2M-Hill, 1985b).

Table 2 DDC impact energies used on test sites near Avenal, California.

Quadrant	Drop Height (m)	Number of primary drops	Number of secondary drops	Total number of drops	Total energy per quadrant (t-m/m²)
1	22.9	10	6	114	385.3
2	18.3	10	6	114	308.2
3	22.9	8	4	88	297.3
4	18.3	8	4	88	238.0

Full-scale dynamic compaction was carried out at the prison in 1986 using a 18.2 t (20 ton) weight dropped from a height of 27.4 m (90 ft). Over 230,000 m² (275,000 yd²) of surface area was treated making this the largest DDC project on collapsible soils in the U.S. No adverse settlement, cracking or distortion has been observed in the prison structure and the dynamic compaction appears to have significantly reduced the potential for collapse settlement.

I-25 between Buffalo and Kaycee, Wyoming

Based on the success of dynamic compaction in Montana, engineering geologists in Wyoming recommended dynamic compaction for treating collapsible soils prior to reconstruction of three large stretches of I-25 between Kaycee and Buffalo Wyoming. Many sections of the highway had become rough and uneven as a result of collapsible soil settlement and maintenance was costly and time consuming. Some sections of the roadway had settled over 0.43 m (17 in) requiring large thicknesses of asphalt overlays to keep the highway level.

The topography in the area consists of rolling terrain with some intermittent drainages. The subgrade materials are Quaternary colluvial and alluvial deposits consisting primarily of silts and sands with some silty gravel layers. The collapsible soil zone was typically 7.6 m (25 ft) thick. Oedometer testing showed collapse strains between 2 to 10%. Moisture contents were typically about 12%.

Two sites, one on the north and the other on the south side of the project area, were selected for field test sites. The test sites were used to determine the effectiveness of DDC on the collapsible soil and the required drop spacing, number of drops, and energy levels. The south test site was divided into 15 separate test grids with each grid consisting of nine primary prints. Each grid had varying parameters and 998 drops were made. The two tamper weights used at the grids weighed 13.6 and 18.2 t (15 and 20 tons). Drop heights of 15.2, 22.9 and 30.5 m (50, 75 and 100 ft) were used along with print spacings of 3, 3.65, and 4.25 m (10, 12, and 14 ft). The number of drops per impact location ranged from 5 to 9. Two of the test grids had secondary prints and cover passes. Cover compaction, dynamic cone penetrometer and standard penetration tests were performed in each grid before and after the compaction at the same testing locations. The dynamic CPT results indicated that the soil resistance increased from 66% to 300% with an average overall soil resistance increase of 184%. Improvement depths ranged from 5.5 m (18 ft) for the lowest applied energy to (9.14 m) (30 ft) for the highest applied energy.

At the north test site 12 test grids were used and a total of 1001 drops were made. Within the test area collapsible soil thicknesses ranged from 4 to 12.2 m (13 to 40 ft). As in the southern test area, both the 13.6 and 18.2 t (15 and 20 ton) weights were used. Print spacings were 3.65, 4.25 and 4.9 m (12, 14, and 16) while the drop heights were 22.9 to 30.5 m (75 to 100 ft). The dynamic CPT showed increases in soil resistance ranging from 50% to 483% with an average overall soil resistance increase of 170%. Crater depths varied from 0.8 to 3.6 m (2.6 ft to 11.8 ft). The large variation in crater depth is partly attributable to the variation in the thickness of the collapsible soil zone as well as variations in the applied energy. The average crater depth was 1.9 m (6.3 ft).

Three separate contracts for dynamic compaction were let between 1988 and 1990. The DDC work was performed using a method specification similar to that used in Montana. In all, over 117,000 m^2 (140,000 yd^2) of surface area were treated. A photograph showing the dynamic compaction crane and a section of the treated highway is presented in Figure 2. The compaction method and expected improvement were based on the test site work described previously. Payment was based on the number of drops and typically ranged from $8 to $9 per drop including

Figure 2 Photograph of dynamic compaction work on I-25 in Wyoming.

mobilization. This translates to a cost of $6 to $7.2 per m² ($5 to $6 per yd²) of surface area treated.

Prior to compaction, the asphalt surface course was stripped but the base course was left in place. The compaction was performed in one pass using a 18.2 t (20 tons) weight dropped from a height of 22.9 m (75 ft). At the primary drop points, located at each corner of a 3 m x 3.65 m (10 ft x 12 ft) grid, six drops were made. An additional two drops were made at a secondary drop point in the center of each grid. Extra drops were required if the last drop increased the depth of the crater more than 10% of the total crater depth. To enforce this specification, crater depths were measured for each of the 84,840 drops. Following the DDC work, the treated area was leveled out with heavy equipment and compacted from the ground surface using conventional compaction equipment. This procedure does not appear to have resulted in complete compaction of the entire surface layer. Any future projects will likely involve alternative surface compaction procedures such as: (1) conventional compaction of the surface layers in lifts following dynamic compaction, or (2) "ironing passes" over the surface with a lighter, larger diameter tamping weight (Hager, Personal Communication, 1993).

Crater depths were typically between 1.2 to 2.4 m (4 and 8 ft) deep. However, crater depths were as high as 4.26 m (14 ft) in a 120 m (400 ft) long section of the highway where the natural moisture content was relatively high. In this section, the deep craters made it difficult to meet the specifications and since it was difficult to extract the weight, the construction process slowed considerably. Subsequent testing indicated that although the craters were much deeper than normal, the soil improvement was not directly proportional. It appears that a soil block was punching into the deep soil layers and causing some heaving to occur which reduced the actual improvement.

I-25 near Bernalillo, New Mexico

In Bernalillo County, New Mexico, collapsible soils are commonly encountered. Interstate 25, comprised of two-lanes in each direction, was constructed through this region and the road surface became distorted due to the collapsing nature of the soils. Based on the New Mexico State Highway Department's previous experience, dynamic compaction was selected for densifying the collapsible soils for a length of 2.2 miles immediately north of the junction with NM 44 on the I-25 alignment. The treatment area is located immediately south of Algodones, and the geology and soil conditions are very similar to those described in the Algodonnes case history.

Bore holes drilled along the alignment showed that the soil profile typically consisted of sandy silts and sands to a depth of 6.1 to 7.6 m (20 to 25 ft) which were underlain by sandy gravel layers. For the sandy silts, the average SPT blow count was 11 and the average moisture content was 9.8%. For the sands and gravels, the

In contrast to most of the projects where dynamic compaction was performed based on a "method specification" (i.e. the number of drops and drop spacing were specified for the contractor) developed from test sections, this project was bid on a "performance specification". The contractor had to ensure that the SPT N value was greater than the following minimum values:

(1) a blow count of 17 or more from 0 - 2.1 m 0 - 7 ft below the treated surface, (2) a blow count of 16 or more from 2.4 - 4.9 m (8 - 16 ft) below the treated surface, and (3) a blow count of 15 or more from 5.2 - 7.6 m (17 - 25 ft) below the treated surface.

Because there were no test sections prior to the production run, the tamper and the grid spacing were modified several times in the initial stages of the project so that the minimum SPT blow count could be achieved. Eventually a 7.93 t (26 tons) steel weight was employed with a drop height of 27.4 m (90 ft). Six drops were made at each corner of a 4.26 m (14 ft) square grid and then at the center of the grid. All post-densification SPT holes were located at the center of the equilateral triangular grid pattern defined by the dynamic compaction imprints. One SPT hole was required for every 186 m^2 (2000 ft^2) of surface area compacted and the minimum N values were achieved in all cases.

No measurements were made of the crater depth as a function of the number of drops at this site since acceptance was based only on final SPT values, however, the final crater depths typically ranged from (0.9 to 1.5 m) 3 to 5 ft in depth (Drumheller, Personal Communication, 1993). Ground vibration was monitored since there was some concern about damage to a gypsum plant adjacent to the highway.

While only two years have elapsed since the compaction was completed, the highway has been performing very well. This may be due in part to the fact that the borrow area between the roadways is asphalted to prevent surface water infiltration and surface water is collected in drain pipes and channeled away from the roadway.

Research Site in Nephi, Utah

Dynamic compaction was performed at the Nephi test site for two different research projects sponsored by NSF. The first project compared the settlement of 1.5 m (5 ft) square footings on collapsible soils after a variety of treatments in comparison with untreated soil (Rollins and Rogers, 1991). The second research study attempted to evaluate the influence of moisture content on compaction efficiency (Rollins et al, 1994). Some of the test pads were pre-wetted to increase the average moisture content.

The test site in Nephi is located near the downstream fringe of an alluvial fan near the base of the Wasatch mountain range in central Utah. The valleys along the southern Wasatch range are in a semi-arid climate which is conducive to the development of alluvial fans where collapsible soils accumulate. Soils in this area generally classify as sandy silts and silty sands but may have a clay content of 5 to 20 percent. The silts and clays in the soil act as a binder which keeps the soil in a loose meta-stable state under dry natural moisture conditions. Some silty gravel layers are interspersed throughout the soil profile. The soil gradation at the Nephi test site typically consisted of 60% silt, 30% sand and 10% silt. The average liquid limit and plasticity index were 25 and 6% respectively. Collapse strains upon wetting at 100 kPa (1 ton/ft^2) varied from 5 to 15%.

Rollins and Rogers (1991) performed dynamic compaction using a 4.1 t (4.5 tons) cubical concrete block dropped from a height of 24.4 m (80 ft). Compaction was attempted on a test cell at the natural moisture content ($w_n \approx 10\%$ and $S \approx 25\%$) and at a test cell which had been pre-wetted to about 70% of saturation. Compaction at the wet site led to liquefaction after a few drops which prevented further improvement. However, at the dry test cell the drop pattern consisted of a square grid spaced at 2.4 (8 ft) on centers with an additional drop in the center of the grid. Compaction was continued until the increase in crater depth was less than 10% of the total crater depth. This criteria led to seven drops at each drop location. The average crater depth was approximately 0.7 m (2.2 ft). The tamper was dropped from a height of 6 m (20 ft) after the first pass to iron the ridges between drop locations.

CPT and PMT testing showed significant improvement to a depth of 3.65 to 4.25 m (12 to 14 ft) following compaction. The pressuremeter moduli and limit pressure showed increases of over 150%. After the dynamic compaction, a 1.5 m (5 ft) square test footing was constructed and loaded to a pressure of 88 kPa (1800 lb/ft^2. Even after wetting to a depth of 3 m (10 ft), the footing settled less than 6.25 mm (0.25 in) contrast with a footing on untreated soil at the site which settled over 0.4 m (15 in).

A second set of dynamic compaction experiments was conducted at the site to evaluate the influence of moisture content on the compaction efficiency (Rollins et al, 1994). Six test cells were prepared, each having a different moisture content. Test cells 5 and 3 had moisture contents of 7% and 10%, respectively and were typical of the variation of the natural moisture content at the site. The average moisture content at the other test cells ranged from 15% to 20%.

A 4.5 t (5-ton) cylindrical concrete weight was used for the dynamic compaction. At each test cell drop locations were centered at the corners of a 2.4 m by 2.4 m (8 ft by 8 ft) square grid. Each drop location received seven drops from a height of about 24.4 m (80 ft). CPT soundings and undisturbed samples showed that the depth of improvement was about 4.25 m (14 ft). In the upper portion of the improvement zone, void ratios decreased from the initial value of about 1.0 to between 0.4 and 0.5. In addition, the potential for collapse was eliminated or greatly reduced within this zone.

Figure 3 presents a plot of crater depths as a function of the number of drops at each of the test cells. The data indicate that the crater depth increases significantly as the average moisture content increases. For the wettest cell the crater depth was nearly 2.4 m (8 ft) deep while at the driest test cell, the crater depth was only 0.46 m (1.5 ft).

This study also found that there was a significant influence of moisture content on compaction efficiency. The results indicate that there is an optimum moisture content for compaction similar to that obtained with conventional Proctor testing and that the optimum moisture content can be predicted using Proctor testing with compactive energies similar to that imparted in the field.

Depth of Improvement

In any dynamic compaction application, it is necessary to estimate the depth of improvement. Depth of improvement is defined as the depth to which the dynamic compaction causes some improvement in a given soil property (i.e. density, stiffness, penetration resistance). Figure 4 shows the depths of improvement from the U.S. case histories along with three data points involving collapsible loessial soils in Bulgaria as reported by Lutenegger (1986) and Minkov and Donachev (1983).

Figure 4 is a plot of depth of improvement, D vs. the square root of the drop energy as first proposed by Menard. Lukas (1986) suggested that 0.5 was a reasonable first approximation for the slope of the best fit line (n value) and listed recommended n values for different soil types. Most of the soils studied in this research are semi-pervious soil deposits, primarily silts, for which n is between 0.4 and 0.5 according to Lukas (1986). The best fit line for the data points in Figure 4 is given by the equation

$$D = 0.40 \, (WH)^{0.5} \qquad (1)$$

where D is depth of improvement in m, W is weight in metric tonnes, and H is drop height in m. Equation (3) is plotted in Figure 4. The n value of 0.40 fits within the range of values suggested by Lukas, however, the coefficient of determination, r^2, is only 0.40.

Slocombe (1993) has suggested that the relationship between depth of improvement and applied energy is non-linear, and that the shape of the curve depends on the initial stiffness or density of the soil. Therefore, a best fit curve was obtained using a second order polynomial equation where depth of improvement was

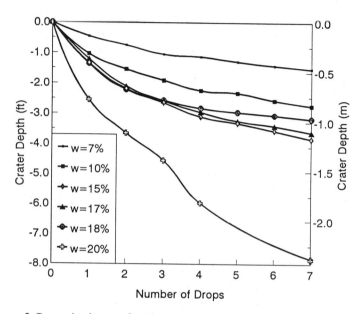

Figure 3 Crater depths as a function of the number of drops for test cells with average moisture contents ranging from 7% to 20%.

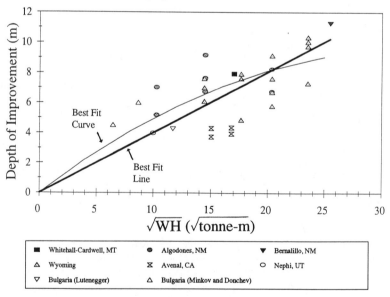

Figure 4 Plot of maximum depth of improvement versus normalized energy for all case histories along with best fit line and best fit curve.

the dependent variable and the square root of the drop energy was the independent variable. The best fit curve is given by the equation

$$D = 0.586 \ (WH)^{0.5} - 0.009 \ (WH) \tag{2}$$

which has an r^2 value of 0.90 indicating high correlation. The best fit curve is also shown in Figure 4. In comparison with the best fit line, the polynomial curve predicts somewhat greater depths of improvement at lower energies but somewhat lower depths at the very highest energies.

Crater Depth

Crater depth measurements after each drop provide an indication of the improvement achieved by the compaction process. Craters are the most visible and immediate result from compaction. Crater depths were measured for most of the case histories described previously.

In order to make comparisons between the various projects which all used different drop heights and weights, the crater depths have been normalized by the square root of the energy per drop. Figure 5 shows the normalized crater depth data from all the case histories along with a typical range defined by Mayne et al (1984) for non-collapsible soils. Most of the normalized crater depths for collapsible soils in their dry natural state plot within the range defined by Mayne et al (1984). Which suggests that their compaction behavior is about the same as non-collapsible soils. In contrast, most of the normalized crater depths for the pre-wetted collapsible soils are greater than for the dry soils and plot below the typical range.

Vibration Attenuation

Ground vibrations caused by dynamic compaction can be damaging to nearby structures and disturbing to people. To ensure that ground vibrations do not exceed acceptable levels, engineers must establish safe distances or compaction energy levels using empirical correlations. Vibration monitoring was performed for the Montana project, the Bernalillo, New Mexico project, and the two Nephi, Utah projects to ensure that vibrations were not damaging to adjacent structures. The frequency content of the vibrations ranged from 5 to 40 Hz.

Vibrations are normally quantified in terms of the peak particle velocity (PPV) which is the maximum velocity recorded in any of the three coordinate axes by a portable seismograph. Well constructed buildings can generally tolerate a PPV of 50 mm/sec, however, a limit of 12.5 mm/sec is often used as a maximum value as a safety margin. To facilitate comparison between various projects, the peak particle velocity has been plotted against the inverse scaled distance as shown in Figure 6. The inverse scale distance is the square root of compactive energy, $(WH)^{0.5}$, divided

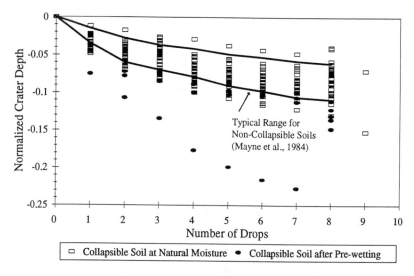

Figure 5 Plot of normalized crater depths versus number of drops for all case histories along with typical range suggested by Mayne et al (1984).

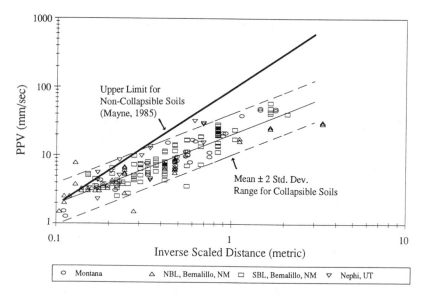

Figure 6 Plot of peak particle velocity (PPV) versus inverse scaled distance for all case history data in comparison with safe upper limit suggested by Mayne (1985).

by the distance, d, from the impact point. The best fit line for the data points is given by

$$PPV = 20(\frac{\sqrt{WH}}{d})^{1.03} \qquad (3)$$

where PPV is in mm/sec, W is m, and H is in metric tonnes. This equation has an r^2 value of 0.82.

The best fit line is shown in Figure 6 along with the $\pm$ 2 standard deviation bounds. The upper limit line defined by Mayne (1985) based on measurements at a large number of sites involving non-collapsible soils is also shown in Figure 6 for comparison. In general, the PPV values for collapsible soils are lower for distances close to the drop point but attenuate (decrease) at a slower rate with distance from the source than for non-collapsible soils. These data suggest that it may be possible to compact somewhat closer to adjacent buildings when performing dynamic compaction on collapsible soils.

Summary and Conclusions

A summary of six case histories involving deep dynamic compaction of collapsible soils in the western United States has been presented and the important compaction parameters for each project are listed in Table 1. Empirical correlations were developed for estimating: depth of improvement, crater depth, and vibration attenuation. These correlations were compared with relationships developed previously for non-collapsible soils.

Based on the case history data the following conclusions are presented:

1. For collapsible soil thicknesses of 6 to 7.6 m (20 to 25 ft), the primary drop spacing has typically been between 1.75 to 2.0 times the tamper diameter and the compactive energy per surface area between 130 and 270 t-m/m^2 (44 and 90 ton-ft/ft^2).

2. The compactive energy per volume used in the various projects has typically been between 12.7 and 44 t-m/m^3 (1.3 and 4.5 ton-ft/ft^3) which corresponds to between 26 and 65 percent of the standard proctor compactive energy.

3. The cost of dynamic compaction, including mobilization, typically was between $6 and $9.5 per m^2 ($5 to $8 per yd^2) of surface area or $0.65 to $1.60 per m^3 ($0.50 to $1.20 per yd^3) of treated soil.

4. The best correlation between depth of improvement and the square root of applied energy is obtained with a polynomial fit. However, for a best fit line, the slope is

0.40 which is within the range of slopes observed for non-collapsible, non-saturated silts and silty sands.

5. Normalized crater depth (NCD) can be related to the number of drops however, the scatter is quite large. Best fit NCD curves for collapsible soils in their dry natural state generally fall within the range defined by Mayne et al (1984) for non-collapsible soils, but normalized crater depths for pre-wetted collapsible soils will be significantly greater than the typical range.

6. Peak particle velocity (PPV) due to DDC correlates well with inverse scaled distance from the drop site. PPV's for collapsible soils are typically lower than those for non-collapsible soils at shorter distances, but the vibrations attenuate more slowly with distance. Vibration frequencies typically lie within the range of 5 to 40 Hz.

References

CH2M-Hill (1985a) "Geotechnical explorations for the California State Prison, Kings County Avenal", Vol. 1, Report M18037.H1, Mar.

CH2M-Hill (1985b) "Dynamic deep compaction test program for the California State Prison near Avenal, Kings County, California, Report M18037.H2, Nov.

Drumheller, J.C. (1993) "Dynamic compaction used in highway construction over landfills and collapsible soils", ASCE National Mtg., New York, Sept.

Lovelace, A.D., Bennett, W.T., Lueck, R.D. (1982) "A test section for the stabilization of collapsible soils on Interstate 25, MB-RR-83-1, New Mexico State Hwy. Dept.

Lukas, R.G. (1986) "Dynamic compaction for highway construction", Vol. 1, FHWA, US DOT, Report No. FHWA/RD-86/133

Luteneger, A.J. (1986) "Dynamic compaction in friable loess." J. Geotech. Engrg., ASCE, 112(6), 663-667

Mayne, P.W., Jones, J.W., and Dumas, J.C. (1984) "Ground response to dynamic compaction" J. of Geotech. Engrg., ASCE, 110(6), 757-774

Mayne, P.W. (1985) "Ground vibrations during dynamic compaction", Vibration Problems in Geotechnical Engineering, ASCE, 247-265.

Minkov, M. and Donchev, P (1983) "Development of heavy tamping of loess bases", 8th European Conf. on SMFE, Helsinki, v. 2, 797-800

Rollins, K.M. and Rogers, G.W. (1991) "Stabilization of collapsible alluvial soils using Dynamic Compaction", Geotech. Special Pub. No. 27, ASCE, 322-333

Rollins, K.M., Jorgenson, S.J., Ross, T. (1994) "Optimum moisture content evaluation for dynamic compaction of collapsible soil", 7th Intl. Congress, Intl. Assoc. of Engrg. Geology, Lisbon, Portugal, (In-Press)

Slocombe, B.C. (1993) Dynamic Compaction in Ground Improvement, Editor, Moseley, M.P., CRC Press, Boca Raton, Florida

Yarger, T.L. (1986) "Dynamic compaction of loose and hydrocompactible soils on Interstate 90 Whitehall-Cardwell, Montana" Trans. Res. Record 1089, 75-80

Acknowledgements

Collection and analysis of the data in this paper was partially supported by NSF Grant BCS-9116339. This support is gratefully acknowledged. The authors are also grateful to Terry Yarger, Montana Dept. of Trans.; Michael Hager, Wyoming Dept. of Trans., Richard Lueck and Edward Rechter, New Mexico State Highway Dept.; Joe Drumheller, Densification, Inc.; Andrew Walker, Geosystems, Inc.; Mark Burgus, Vibra-Tech, Inc; and George Sisuentes, Calif. Dept. of Corrections who graciously opened their files and provided background information for the projects described in this paper.

Evaluation of Dynamic Compaction of Low Level Waste Burial Trenches Containing B-25 Boxes

Scott R. McMullin[1]

Abstract

The Savannah River Site, owned by the U.S. Department of Energy, is preparing to close an additional 13.8 ha of burial grounds under the Resource Conservation Recovery Act. In preparation for this closure, the dynamic compaction facility was designed and constructed to address unresolved design issues. Among these issues is the evaluation of the ability for dynamic compaction to consolidate buried low level waste containers. A model burial trench containing simulated clean wastes was dynamically compacted, after which the materials were excavated and compaction quantified.

The test determined that under existing success criteria, the bottom tier of stacked B-25 boxes were not being consolidated. A quasi-structural layer was formed midway through the stacked boxes, which absorbed the compactive energy. Resulting from these observations and the data collected, a new success criterion is recommended which depends on the relative displacement per drop. The test successfully demonstrated that dynamic compaction will consolidate buried metal boxes.

Introduction

The Savannah River Site (SRS) is a U.S. Department of Energy (DOE) production facility located in western South Carolina. In operation since the early 1950's, this facility's production processes have generated a variety of wastes. Low level radioactive wastes were deposited in the SRS burial grounds, located near the site's geographic center. Solvent wipe rags erroneously deposited with these low level wastes, forced a settlement agreement between the DOE and the South Carolina Department of Health and Environmental Control. This settlement agreement mandates burial ground closure as a mixed waste facility under the rules and regulations of the Resource Conservation Recovery Act. SRS completed the

---

[1]Senior Geotechnical Engineer, Environmental Restoration Engineering, Westinghouse Savannah River Company, P.O. Box 616, Bldg. 992-4W, Aiken SC 29803

initial 23.5 ha closure of the Mixed Waste Management Facility (MWMF) in 1991 (Schexnayder and Lukas, 1992). Preparations are currently underway to close an additional 13.8 ha in the future.

Current disposal procedures require placement of low level wastes in 1.2 x 1.2 x 1.8 m low carbon steel containers called B-25 boxes. The boxes are stacked four high in Engineered Low Level Trenches (ELLT), optimizing burial space. Some SRS ELLTs contain upwards of 20,000 boxes. After placement, the ELLT is backfilled and covered with 1-2 m locally available of sand to sandy clay soil.

Prior to closure plan development and the associated closure, the SRS identified several issues needing resolution. Among these issues and the focus for this paper is the evaluation of the effectiveness of dynamic compaction to consolidate buried low level radioactive wastes. This evaluation includes validation of the SRS success criterion and a quantification of void reduction. To resolve these issues, the dynamic compaction facility was designed, constructed, and tested (Lukas, 1986). This paper will first present a brief overview of the dynamic compaction test, followed by a discussion of the key observations, the test results, and the conclusions.

Test Design/Instrumentation

The dynamic compaction test facility is a model ELLT, containing 168 B-25 boxes filled with simulated low level radioactive wastes. These wastes consist of four basic categories: metal, wood, soil, and simulated personal protective equipment. The boxes were stacked in a 7x6x4-box matrix, with a unique identifying number assigned to each box. During backfill, care was taken to assure intimate contact with the soil.

Vibratory Ground Motion Monitoring Program
An initial assumption for the dynamic compaction facility test, was that as the buried wastes approached maximum density, the attenuation of the peak particle velocity (PPV) would approach a characteristic attenuation curve for natural, undisturbed soils. To monitor PPV, an array of 43 strong motion sensors was used during the dynamic compaction facility testing. PPV was measured in the x, y, and z axes for each dynamic compaction weight drop at each instrument location.

Measurement of Impact Velocity
Dynamic compaction is done by hoisting a weight to a specified height, then releasing the weight to impact with the target material. Every crane has an inherent efficiency factor, which depends on the internal friction for the cable spool, brake configuration, and associated pulleys. Production cranes are designed to minimize this internal friction to maximize the impact energy. These cranes typically attain an efficiency factor ranging between 80% - 90% (Mayne, 1985).

To realistically model actual closure conditions, an efficiency for the dynamic compaction crane was measured and differences calibrated against a production type crane. The efficiency factor is determined by measuring the impact velocity and comparing the kinetic energy to potential energy, as shown below.

Kinetic Energy:

$$E_k = \frac{1}{2}mv^2 \qquad (1)$$

Potential Energy:

$$E_p = mgh \qquad (2)$$

Combining equations (1) and (2) to compute the efficiency factor:

$$\text{Eff} = \frac{E_k}{E_p} = \frac{v^2}{2gh} \qquad (3)$$

where
  Eff = efficiency factor
  E_k = kinetic energy at impact,
  E_p = potential energy,
  m = mass of the weight,
  v = measured velocity at impact,
  g = acceleration due to gravity,
  h = drop height.

The critical measurement for calculating the efficiency factor is the velocity at impact. Instrumentation used during this study to measure velocity at impact consists of a series of laser diode photoelectric cells mounted along two vertical poles at predetermined heights. As the weight falls downward, it sequentially brakes a series of photocell beams at several predetermined heights above the ground surface. Since the photocell beams are located a known distance apart, the interval velocity and interval crane efficiency can be calculated. Determine instantaneous impact velocity is then a simple matter of extrapolation using the quantities measured in the photocell gates.

The efficiency factor for the dynamic compaction crane was determined to be 55.5%, with a standard deviation of 1.5%. This value is different from the 80%-90% for a production crane and is attributed to differences in friction, brake, and cable configuration. A production crane is designed to optimize the performance of these factors to maximize the drop energy. The crane used for this test was design for lifting, thus the difference in crane efficiency.

Test Criterion
The test criterion for the dynamic compaction facility test modeled the criterion determined for the MWMF closure. The MWMF closure specified a 18.1-t dynamic compaction weight dropped from 12.8 m. The success criterion was a 1.8 m depression or 20 consecutive drops, whichever came first. Based on the evaluation of crane efficiencies, the test drop height was increased from 12.8 m to 15.2 m. This change provided compactive energy similar to that anticipated during actual production.

Stress Influence with depth

Preliminary evaluation of the anticipated influence of stress with depth suggested that the traditional success criteria may not effectively consolidate the stacked B-25 boxes. Work has been conducted on determination of impact stresses with depth in various soil types (Mayne and Jones, 1983). Simple equations to calculate impact stresses require determination of controlling soil parameters. Determination of these parameters is precluded at SRS because of the heterogeneities present in the SRS buried wastes. The Boussinesq equation is material independent. With simplifying assumptions, application of the Boussinesq Equation provided an estimate of the vertical stress influence with depth.

The instantaneous velocity for each impact was measured as part of the dynamic compaction test program. Knowing this velocity at impact, the deceleration is computed using equation (4):

$$v = v_o + 2a(y_1 - y_o) \tag{4}$$

where
- $(y_1 - y_o)$ = differential displacement induced by weight impact
- a = weight deceleration
- v_o = velocity of the weight at initial impact
- v = final weight velocity ($= 0$)

Knowing the deceleration, the stress induced by the impacting mass can be calculated using:

$$q = \frac{ma}{\pi r^2} \tag{5}$$

where
- q = induced vertical stress,
- r = radius of dynamic compaction weight.

Combining equations 4 and 5 to estimate the stress induced at the surface, the value is entered into the Boussinesq Equation for a circular load to quantify the induced stress with depth (Das, 1985).

$$\Delta p = q \left\{ 1 - \frac{1}{\left[\left(\frac{r}{z}\right)^2 + 1\right]^{\frac{3}{2}}} \right\} \tag{6}$$

where
- Δp = change in vertical stress
- z = depth of influence for which Δp is calculated

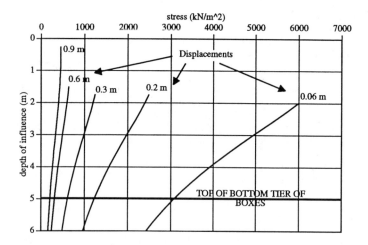

Figure 1. Plot of the influence of stress with depth, using the Boussinesq equation.

Figure 1 presents the Boussinesq equation as a plot of stress versus depth of influence as a function of the differential displacements by the impacting weight. A 0.9 m displacement relates to an average first drop in a primary pattern, while a 0.06 m change relates to a late secondary/tertiary type drop. Assuming the bottom tier of boxes is 5 m deep, the 0.9 m displacement inducers very little additional vertical stresses in the bottom boxes. Theoretically, as the box matrix becomes more stiff, the differential displacement becomes smaller, increasing stress with depth. At the 0.06 m displacement range the bottom tier of boxes is more likely to be influenced.

Though the Boussinesq equation is material independent, the intended application is to stresses induced to a soil matrix. As this application is to metal boxes, the calculated results should be calibrated to assure applicability. There were no stress measurements taken at depth to provided this calibration. A useful evaluation can be deduced by comparing to a model developed by Slocombe (Slocombe, 1993). Slocombe presents a model of depth of influence as a function of scaled energy. The average kinetic energy at impact for this test is 16.6 t-m, at the computed differential displacement ranging between 0.9 m and 0.06 m. Superimposing this value over the range of influence plotted on the Slocombe graph, the range of data fits rather well over the data generated for soils (figure 2). With this correlation, it is assumed that applying the Boussinesq equation provides a reasonable estimation of the depth of influence for the SRS dynamic compaction efforts.

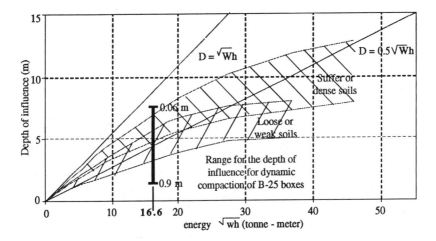

Figure 2. Plot of scaled energy versus depth, compared with Slocombe correlations for soil. (after Slocombe)

Based upon the preliminary evaluation, the drop pattern for the dynamic compaction test was modified to assess the SRS traditional success criteria. The drop pattern for the test was modified to consist of two drop zones (figure 3). Drop zone A, was compacted to the traditional SRS success criterion, while drop zone B was over compacted using an additional tertiary drop pattern.

Preliminary Attenuation Baseline

A drop test on natural, undisturbed soil (dense SC soils) was done to establish a baseline strong motion attenuation curve. An array of strong motion sensors was placed around the drop location, recording the PPV for the shear wave energy. The shear waves were generated by successively dropping the weight from varying heights ranging between 3.0 and 15.2 m. The initial

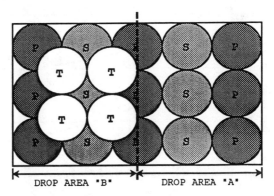

Figure 3. Drop area A, compacted to existing specifications; drop area B, over compacted to facilitate generation of a new specification.

assumption was that as the buried wastes reached higher densities, the attenuation curve for the buried wastes would approach the baseline for natural soil. Comparison of these two curves allows an evaluation of the success of the dynamic compaction during consolidation (Dobry and Gazetas, 1987; Woods and Jedele, 1987).

Observations

During dynamic compaction, there were several observations leading to the conclusion that the traditional SRS success criterion was not adequate and that the buried boxes were not being effectively consolidated. The first observation was a difference in the shear wave energy moving through the soil. During preliminary drops on natural soil , the ground motion was noticeable. During the first series of drops on the buried B-25 boxes, there were no noticeable shear waves felt at a distance of 15 m. The buried box matrix appeared to absorb the weight impact without transmitting shear wave energy to the soil (Haskel, 1953).

The second observation was that the attenuation curves for the PPVs were not approaching the baseline curve (figure 4). At a distance of 3 to 24 1/2 m from the drop point, the PPV ranged from 15.2 to 0.3 cm/s, respectively for the first drop location. As drops continued through the primary, secondary, and then tertiary drop patterns, there was very little variation in the PPV. There were 128 drops during the test, with the last fourteen at the same tertiary location. At distances from

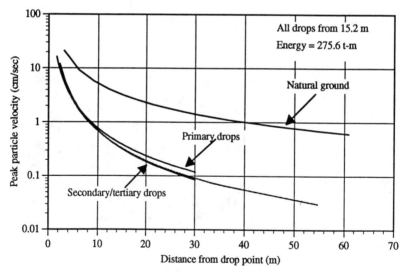

Figure 4. Semi-log plot comparing attenuation of peak particle velocity versus distance between the natural ground baseline and the primary, secondary, and tertiary drops.

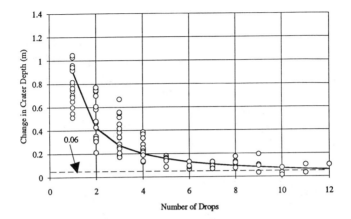

Figure 5. Plot comparing the number of drops with respect to the change in crater depth.

4 to 24 1/2 m, the PPV ranged between 16.4 and 0.5 cm/s, respectively. There was very little difference between the PPVs measured on the initial drops and those measured on subsequent drops.

The third observation, was that the dynamic compaction weight "bounced" several times upon impact. A linear scale placed on the side of the weight was photographed using high speed video photography. Examination of the high speed video photography revealed that the bounce effect increased with the number of drops. An explanation was formulated that an "elastic layer" of consolidated boxes was formed during compaction. It is presumed that as the boxes consolidated, they were interlocking creating a quasi-structural layer.

The final observation was that the traditional, SRS success criteria was not allowing full consolidation of the buried boxes. On the very first drop, the weight easily achieved the 1.8 m displacement criterion. Based on observation and preliminary calculations using the Boussinesq equation, there was little densification occurring for the lower boxes. The change in crater depth per drop, for the primary drop pattern ranged between 0.6 m and 1.5 m. For secondary drop sties, the number of drops required to reach 1.8 m ranged between four and six drops. Only in the tertiary drop patterns on side B, did the change in crater depth per drop become relatively small. Figure 5 is a plot comparing the drops with respect to the change in crater depth. The curve is a best fit for all drop data. Note that this curve is becoming asymptotic near an approximate change in crater depth of 0.06 m.

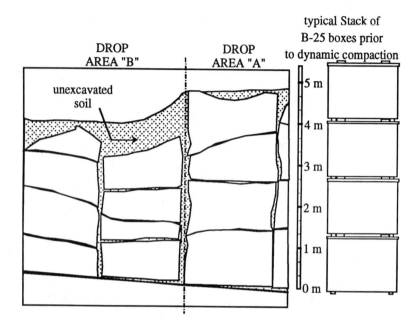

Figure 6. Westerly edge of excavated B-25 boxes, showing the difference between compaction zones A and B. Note that sketch is digitized from an actual photograph.

Buried Waste Compaction Evaluation

Excavation of the consolidated B-25 boxes showed that as suspected, dynamic compaction did not successfully consolidate the lower tiers of boxes, except in areas of over compaction (based on the SRS success criterion). The boxes were excavated one by one, with measurements made of each dimension. The volume was compared with the initial volume to determine the percent compaction. Figure 6 is a digitized photograph of the western side of the box matrix, with a sketch of the original configuration added for reference. Note that some boxes in drop area A show little consolidation, while the boxes in drop area B are much more compressed. Generally, boxes in drop area B were 30% more compacted than those in drop are A, with local variations. This information supports the initial suppositions that the traditional SRS criteria was not successful.

Excavation also revealed that the compacted boxes were interlocking forming a quasi-structural layer. This layer was apparently formed by the lateral spread and interlocking of the compacted boxes. Failed boxes and materials overlaid each other so tightly that cranes extracting the boxes tore the metal rather than separating the boxes. Additionally, the outside edges of the box matrix were not effectively consolidated. Apparently, the drop pattern did not overlap the exterior edge of the box matrix. The drop patterns were designed so that the distance between the center point of impact and the exterior edge was the radius of the weight. Failure to consolidate the box matrix edge may have contributed to the elastic nature of the interlocked, box layer by providing supports to bridge the box matrix.

The design purpose for B-25 boxes is to contain low-level waste at the generation point, to protect workers, and to facilitate transportation to the burial site. These boxes were never intended to contain waste within a burial trench; however, they do by default help minimize waste migration. An excavation of the consolidated boxes revealed that dynamic compaction accelerated box degradation by corrosion. As part of a four-year corrosion study, three B-25 boxes were buried uncompacted at a nearby location. When the boxes were excavated, the Savannah River Technology Center concluded that these undisturbed boxes experienced no observable degradation from corrosion. The dynamic compaction test boxes had been in the ground less than six months when excavated. Corrosion of the dynamic compaction test boxes exceeded that of the long-term study, demonstrating that dynamic compaction had accelerated box degradation due to corrosion.

Conclusions

The dynamic compaction test at the Savannah River Site successfully addressed the test objectives. These objectives included an evaluation of the effectiveness of dynamic compaction to consolidate buried low level wastes and it is concluded that:

1. The use of strong motion sensors as a tool to monitor the success of dynamic compaction was only partially successful. This tool did provide early evidence that the SRS test was not compacting the lower tiers of boxes. However, the metal boxes absorbed the compact energy, reducing and distorting the wave forms transmitted through the soil. This distortion precluded the successful use of strong motion sensors to quantify the densification of the metal boxes during dynamic compaction.

2. Excavation of the compacted wastes to evaluate the dynamic compaction success provided valuable insight that will improve the quality of future closures. The observation of the need to improve the SRS success criterion and the formation of an interlocked, elastic layer are both new and unique. The traditional SRS success criteria was a 1.8 m displacement or 20 consecutive drops, whichever came first. The observations early in the test that these criteria were inadequate allowed modifications to the test plan and the success of the program.

3. As the boxes were excavated, the measured reduction in void ratio for the traditional criteria demonstrated the general lack of compaction to the bottom tier of boxes. Excavation of the over compacted boxers demonstrated the value of the new SRS criterion of two consecutive drops with a change in crater depth of no greater than 0.06 m. Two consecutive drops at this

criteria will ensure that no minor bridging of material occurs at the drop site. If the subsurface material bridge can withstand the energy imparted by two consecutive drops, then additional drops to destroy the bridge will most likely not be cost-effective.

Appendix. References

Das, B.M.. (1985). *Principles of Geotechnical Engineering,* Prindle, Weber & Schmidt (PWS) Engineering, Boston, MA.

Dobry, R. and G. Gazetas. (1985). "Dynamic Stiffness and Damping of Foundations by Simple Methods." *Vibration Problems in Geotechnical Engineering,* (Proceedings, ASCE Convention, Detroit) American Society of Civil Engineers, New York, pp. 75-107.

Mayne, P.W. (1985). "Ground Vibrations During Dynamic Compaction." *Vibration Problems in Geotechnical Engineering,* (Proceedings, ASCE Convention, Detroit) American Society of Civil Engineers, New York, pp. 247-265.

Haskel, N.A.. (1953). "The Dispersion Of Surface Waves In Multilayered Media," *Bulletin of the Seismological Society Of America,* Vol. 43, pp. 17-34.

Lukas, R.G. (1986), "Dynamic compaction for Highway Construction: Design and Construction Guidelines," *Report FHWA/RD-86/133,* Federal Highway Administration, Washington, D.C., 230 p.

Mayne, P.W. and J.S. Jones, Jr. (1983), "Impact Stresses during dynamic compaction," ASCE *Journal of Geotechnical Engineering.* 109(10), pp. 1342-1346.

Slocombe, B.C. (1993). "Dynamic Compaction," *Ground Improvement,* CRC Press, Inc., Boca Raton, Florida, pp. 20-40.

Schexnayder, C. and Lukas, R.G. (1992), "The Use of Dynamic compaction to Consolidate Nuclear Waste", *Grouting, Soil Improvement and Geosynthetics,* (Proceedings, ASCE and ISSMFE special conference) American Society of Civil Engineers, Geotechnical Special Publication No. 30, pp. 1311-1323

Woods, Richard D. and Larry P. Jedele. (1985). "Energy-Attenuation Relationships From Construction Vibrations." *Vibration Problems in Geotechnical Engineering,* (Proceedings, ASCE Convention, Detroit) American Society of Civil Engineers, New York, pp. 229-246.

The information contained in this article was developed during the course of work under Contract No. DE-AC09-89SR18035 with the U.S. Department of Energy. By acceptance of this paper, the publisher and/or recipient acknowledges the U.S. Government's right to retain a nonexclusive, royalty-free license in and to any copyright covering this paper along with the right to reproduce, and to authorize others to reproduce all or part of the copyrighted paper.

Dynamic Compaction: Two Case Histories
Utilizing Innovative Techniques

Albert A. Bayuk[1] and Andrew D. Walker[1], M. ASCE

Abstract

To keep foundation costs to a minimum, weak or poor ground conditions are frequently improved at depth by dynamic compaction. These weak deposits may be natural, but are often man-made. Two recent case histories are reviewed where dynamic compaction was used to improve ground support of man-made/natural deposits for:

1. A one-story retail building, containing approximately 11,000 square meters of conventional spread footing and floor slab on-grade construction; and

2. Approximately 80,000 square meters of planned landfill expansion at a power plant, whose height of residual waste is expected to approach 43 meters.

The two case histories presented are noteworthy since innovative variations to the basic dynamic compaction technique were introduced at both sites.

CASE 1 - RETAIL BUILDING

Introduction

The subsurface profile for The Home Depot one-story retail building and adjacent parking structure is typical of the low-lying meadowlands in Secaucus, New Jersey. Basically, the site is blanketed

---
[1]Project Manager, Geo-Con, Inc., 4075 Monroeville Boulevard, Monroeville, PA 15146

by 3 to 4.5 meters of uncontrolled miscellaneous granular fill, which is underlain by approximately 1 to 2 meters of compressible organic soils. Beneath the fill/organic soils is a thick deposit of varved silts and clays. Project documents specified the most economical ground improvement technique: dynamic compaction, to improve the fill/organic materials, and thus permit the use of conventional spread foundations and floor slab on-grade construction. The intent was to force the lower portions of the miscellaneous fill (containing large concrete pieces, rock fragments and building rubble) down into, and mix with the organic soils to achieve improvement.

Experience has shown that achieving sufficient energy transfer by dynamic compaction to cause adequate intermixing of weak with stronger materials is marginal at depths similar to the Secaucus site. For this reason, Geo-Con could not guarantee compliance with the specification, and proposed that the specified test section be evaluated in advance of any production work.

Total and differential settlements of the test section were unacceptable for the heavier proposed building footings and floor loadings, and a second test section was performed utilizing a combination of dynamic compaction and stone columns. Based on the performance of test section nos. 1 and 2, the client's design team determined how each of the project structures were to be supported.

<u>Generalized Subsurface Profile</u>

Subsurface conditions at the site were reported by the project's geotechnical engineer (Melick-Tully and Associates, Inc., 1990) and are reasonably uniform. The generalized subsurface profile is shown in Figure 1.

Groundwater was encountered in all borings at the site, ranging from 0.6 to 3.7 meters below existing ground surface.

<u>Building Areas and Loadings</u>

The one-story retail building is approximately 11,000 square meters in plan area, including the attached garden center. The structure is a steel frame with pre-cast concrete exterior walls, and uniformly spaced interior columns. Maximum continuous wall loadings are 65 kN/m and interior column loads are typically 360 kN. Generally, the design floor slab live load is 12 kN/m^2, with about one-third of the building designed for 30 kN/m^2.

A one-story parking structure is located adjacent to the retail building, and is approximately 8,000 square meters in area. Maximum exterior and interior column loads are on the order of 445 kN and 1330 kN, respectively.

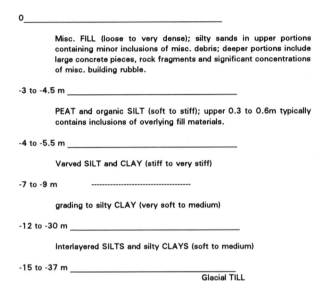

Figure 1. Generalized Subsurface Profile

Test Section No. 1

Test section no. 1, whose impact print and soil boring plan are shown in Figure 2, demonstrated that dynamic compaction produced a high degree of improvement within the miscellaneous granular fill overlying the site. A plot of the average degree of improvement within the upper fill is shown in Figure 3, and indicates that ground improvement after dynamic compaction ranged between 150 to 260%.

Surcharge loading and settlement plate locations are shown in Figure 4. Subsequent monitoring of the settlement plates indicated that differential movements and angular distortions were excessive for the test surcharge loading on the underlying peat. Only for small lightly loaded footings, whose significant bulb of stress could be contained within the highly densified miscellaneous granular fill, would resulting differential movements and angular distortions be acceptable.

Based on the performance of test section no. 1, it was concluded that dynamic compaction was unable to drive the upper granular fill down into the peat and organic silts and, therefore, would not achieve the required results.

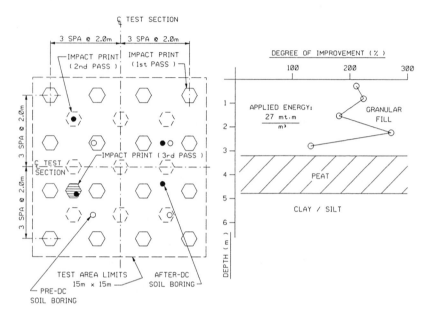

Figure 2. Test Section Plan   Figure 3. Degree of Improvement

Test Section No. 2

To insure a more positive intermixing of less compressible granular material with the underlying weak organic soils, stone columns were installed and dynamic compaction performed at a second test section. In addition to installing a stone column at each dynamic compaction print location, the print spacing was reduced from 4m to 3.7m and the number of drops at each print increased from 6 to 8, resulting in a 50% increase in applied energy above that used at test section no. 1. The load test at test section no. 2 was conducted in a similar manner to that performed at test section no. 1, and under the same surcharge loading.

Monitoring of the settlement plates indicated approximately a 20% reduction (nearly 13 mm) at test section no. 2 in average total settlement after a similar time duration (16 days). Most significant was the low differential movement (less than 13 mm) at test section no. 2

between settlement plates installed over similar stone column configurations.

Selection of Ground Improvement Schemes

Based on the performance of test section nos. 1 and 2, the client's design team determined which areas of the project could be supported on conventional spread foundations and slab on-grade construction after ground improvement by dynamic compaction or dynamic compaction/stone columns.

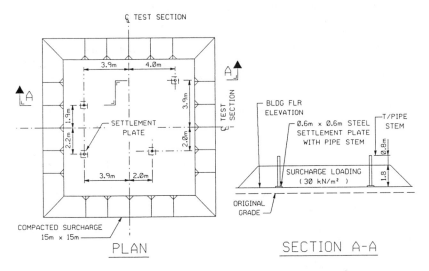

Figure 4. Surcharge Loading and Settlement Plate Locations at Test Section

Due to the more heavily loaded column footings at the parking structure, it was decided to support this structure on piled foundations.

Stone Column Installation

At the more heavily loaded floor areas and more settlement sensitive footings, stone columns were installed prior to performing dynamic compaction at the retail building. A total of 765 stone columns were installed, including those added due to occasional relocations at difficult penetration areas after encountering obstructions within the miscellaneous granular fill.

The stone columns were installed utilizing the wet top-feed method and Geo-Con's powerful depth vibrators. Backfill was a crushed angular stone ranging in size from 19 mm to 75 mm. Approximately 3200 metric tons were placed in 765 stone columns that averaged nearly 4.3 m in depth. It is estimated that at least one-half of the total amount of crushed stone used was consumed in the lower one-third of the stone column depth. Therefore, the average stone consumption rate within the soft organic soils was at least 1.5 metric tons per meter, and formed about a 1.1 m diameter stone column.

Dynamic Compaction Methods and Techniques

Dynamic compaction was performed at all areas of the retail building. Some areas received only dynamic compaction at applied energies consistent with those used at test section no. 1, while other areas received a combination of both dynamic compaction/stone columns consistent with test section no. 2. All high energy passes were performed using an 11.8 metric ton tamper weight and a drop height of 16.8 m.

Tamper impact craters were filled by collapsing the ground between craters and also pushing on-site stockpiles of select granular material into the cavities. The site was leveled between passes and tracked with a dozer. A low energy ironing pass compacted the crater materials and upper portion of the miscellaneous fill.

Conclusion

The soils profile at The Home Depot site precluded exclusive adherence to the project specifications. Improvement within the 1 to 2 m of compressible organic soils underlying the site at a depth of about 3 to 3.5 m could not be guaranteed, and required the evaluation of two test sections prior to production dropping of the high energy tamping weight.

The initial test section performance concluded that dynamic compaction, as specified, could only be used effectively at small footings and lightly loaded floor slabs. A second test section, utilizing stone columns to complement dynamic compaction, showed that this innovative combination of ground improvement techniques would permit the use of conventional spread foundations and slab on-grade construction even at the more heavily loaded areas of the retail building.

The unique combination of dynamically compacted stone columns permitted the use of a shallow foundation system on a site with a very difficult soils profile, and at significant cost savings.

CASE 2 - LANDFILL EXPANSION

Introduction

When Columbus Southern Power's Conesville power station needed additional space to store scrubber waste, management decided to extend its existing 43 m high landfill. The expansion placed the new landfill directly over an old strip mining site, that formerly had supplied the power station with its coal.

The new landfill expansion, encompassing an area of approximately 80,000 square meters, is primarily underlain by two types of mine spoil. The northern portion of the area consists of weathered sandstone and shale shotrock piles, formed when the valley was stripmined in the early 1960's. At the southern portion of the site, the mine spoil consists of the overburden soils removed to expose the bedrock prior to blasting and rock excavation.

A major technical requirement was to reduce the potential for the landfill expansion to induce large differential settlements in localized areas of the underlying mine spoil. Such movements would severely impact the PVC and clay liner underlying the landfill, and therefore affect the groundwater. Dynamic compaction was utilized to improve the heterogeneous nature of the mine spoils, so as to produce a more uniform soil response when the landfill is expanded, thus limiting localized differential settlements.

Measuring and monitoring penetration of the tamper weight into the mine spoil for approximately 30,000 high energy drops were critical control parameters for normalization of the subsoils. An automatic measuring/recording device was successfully developed and installed on each crane's cable drum. This innovative instrumentation was an essential component of the quality assurance/quality control program at the site.

Description of Mine Spoil

The rock mine spoil at the northern portion of the site consists of large rock fragments within a decomposed rock matrix of weathered shale and sandstone. The southern deposits are more variable; consisting of soft to stiff silts and clays, and loose to dense sands with varying amounts of rock fragments. The soil mine spoil, also referred to as overburden spoil, is essentially a cohesive deposit with medium plasticity clays predominating.

Based on information from mining maps obtained from the client (Slomski, Stephen, 1993), thicknesses of the mine spoil ranged from less than 3 m to more than 9 m. Typically, rock spoil is 7.6 to 9.2 m thick and overburden spoil about 6 m. In general, groundwater is located near the base of the mine spoil deposits.

Test Sections

Basically, two test sections were performed within the dynamic compaction limits and established production parameters. One test section was located over the rock spoils, and the other within the overburden spoils.

Important dynamic compaction parameters were monitored at each test section, and included the following:

1. tamper penetration rate,
2. crater expansion,
3. volume of ground heave,
4. ground subsidence,
5. vibration monitoring, and
6. depth of improvement.

Due to the heterogeneous nature of the mine spoil, along with its fine-grained residual and depositional structure, before and after dynamic compaction testing revealed inconsistent improvement in the standard penetration test values. As a result, depth of improvement could not be ascertained with any degree of certainty.

The test program reinforced the owner's proposed concept that the principle quality control/quality assurance parameter was the rate of penetration of the tamper weight into the mine spoil. Ground subsidence, as a percentage of the original mine spoil thickness, measured dynamic compaction performance.

Grid patterns, drop heights, and rates of crater expansion were determined to produce the necessary ground response at the mine rock and overburden spoils so as to normalize their respective stiffnesses.

Grid Pattern and Drop Heights

The site was divided into 30.5 m quadrants, encompassing the rock spoils at the northern sections and the overburden spoils to the south. A split spacing of 4.3 m centers, initially installed in two passes, was used at the overburden spoils. A single pass at 4.3 m centers was proposed for the northern rock spoils. However, the test program revealed a higher degree of fine-grained material within the rock spoils than was anticipated. As a result, the same spacing and passes installed at the overburden spoils were also used at the rock spoils during production work.

A 19.1 metric ton tamping weight was dropped from heights ranging between 6 and 21 m, depending on the estimated mine spoil thickness. Drop heights were determined by utilizing the relationship proposed by Menard and Broise (Menard, L. and Broise, Y., 1975) and the suggested modification presented in the FHWA Report (Federal Highway Administration, 1986) as follows:

$$D = n(WH)^{1/2}$$

where:

D = depth of improvement in meters
W = weight of tamper in metric tons
H = drop height in meters
n = empirical coefficient (less than 1)

Rate of Crater Expansion

During production work, the tamper rate of penetration into the mine spoil was the variable parameter. It was the principle measuring/monitoring observation used to assess the relative stiffness of the mine spoil at each tamper impact location.

During the early test program, drop heights of 15 m were found to exhibit efficient crater volume/heave ratios at both the rock and, particularly, the softer overburden spoils. Generally, tamper penetration was measured after each drop, and measurements were always attempted during the final three drops. Plots of drop number versus average tamper penetration are presented for the rock spoils and the overburden spoils in Figure 5.

Six drops were initially proposed for the overburden spoils, and crater expansion (i.e. resulting crater volume due to tamper impacts) for this number was extracted from the field data and is summarized below:

Average Crater Expansion at Overburden Spoils
(6 drops from 15 m)

Depth (m)	Diameter (m)		Volume (cu. m.)
	Top	Bottom	
1.90	2.90	2.14	9.46

Crater depths exceeding about 1.8 m are difficult for the tamper weight to re-enter without striking the sides of the crater, especially from drop heights of 15 to 21 m. Also, proper backfilling and compaction of the craters become difficult. As a result, a maximum of six drops per pass was selected in the overburden spoils. Referring to Figure 5, the average tamper penetration at the fifth drop is 0.25 m and at the sixth drop is about 0.19 m.

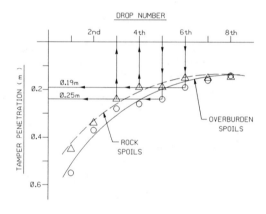

Figure 5. Plots of Drop Number Versus Tamper Penetration

Referring to Figure 5 again, similar rates of tamper penetration for the rock spoils are noted at the third and fourth drops. Crater expansion for four drops from 15 m at the rock spoils is summarized as follows:

Average Crater Expansion at Rock Spoils
(4 drops from 15 m)

Depth (m)	Diameter (m)		Volume (cu. m.)
	Top	Bottom	
1.31	2.87	2.14	6.46

As expected, less drops and smaller volume changes were achieved in the stiffer rock spoils for the same tamper penetration rates. Volume change was directly proportional to applied energy; both being approximately 50% greater at the overburden spoils.

Rate of tamper penetration, together with crater depth, was used to establish the criterion for termination of high energy drops at over eight thousand impact print locations. An impact print was terminated when one of the following conditions was achieved:

1. rate of tamper penetration totaled 0.6 m or less for two consecutive drops; or
2. crater depth exceeded 1.8 m.

If the crater depths exceeded 1.8 m before the rate of tamper penetration was achieved during the secondary pass, the craters were backfilled and additional passes performed until the tamper penetration criterion was established. It is significant to note that all quadrants requiring more than two passes were located within the overburden spoils, or at the transition zone between the overburden and rock spoils. As many as five passes, totalling 15 drops, were required at some of the softer overburden spoils.

Automated QA/QC Monitoring

Not only did the northern and southern portions of the mine spoil vary due to their different geologic structures, but their unnatural depositions resulted in varying stiffnesses within relatively short distances of their own profiles. One of the merits of the dynamic compaction technique is its inherent stiffness testing feature. Weak and strong areas are identified by how the soil responds to the dropped weight. Dynamic compaction can be easily adjusted in the field to apply more energy at the weaker areas. Using a tamper impact print spacing of 4.3 m centers in a split pattern, this self-proving feature provided a quality assurance test for every 9.2 square meters.

Crater expansion, particularly tamper penetration rate, was the principle quality assurance/quality control parameter for production work. Measurement of tamper penetration for each drop, crater depth and number of drops for over eight thousand high energy impact prints generated an enormous amount of data.

To effectively and efficiently record and process this daily influx of data, a purposely designed tachometer was installed on each crane's cable drum and is shown in Figure 6. This information was transposed to a digital readout of the elevation of the weight, before and after each drop, which was displayed in the crane cab. Thus, the operator knew immediately when the criteria for terminating a print had been achieved. In addition, a hard copy of this data was provided in graphical form on a strip chart recorder for each tamper drop.

Crater expansion data was uniquely summarized for each quadrant by GAI Consultants, Inc., who monitored the quality assurance/quality control program. These quadrant summary sheets pictorially illustrated location, total drops, crater depth, and tamper penetration for the final two drops at each impact print location. Summary presentations in this form were very helpful during the evaluation of the QA/QC data. Weak areas of each quadrant were easily located and additional tamper drops performed to increase their stiffness.

Conclusion

Varying geologic structure and unnatural deposition of mine spoil underlying Columbus Southern Power's proposed landfill expansion was

Figure 6. Tachometer Mounted on Crane Cable Drum

a major concern of the Ohio Environmental Protection Agency. Before the EPA would approve the expansion, the owner had to submit a plan for reducing the potential for differential settlements within the mine spoil. At a cost savings of $4 million, dynamic compaction was chosen over conventional undercut and replacement methods.

Soil sampling and testing, site excavations, ground response to the tamper weight, and ground subsidence measurements, confirmed the Ohio EPA's concerns about the heterogeneous structure of the mine spoil. The number of drops at each impact print to achieve the tamper penetration criterion ranged between 2 and 15. Average quadrant induced settlement, as a percentage of mine spoil thickness, ranged between 2 and 11%.

By the use of fully instrumented cranes supported by an automated data collection system, a sophisticated quality assurance/quality control program was implemented. The direction and control afforded by this system was essential to ensure a consistent final ground response to the impact of the tamper weight, and thus a more uniform soil stiffness following compaction.

The real-time data generated, amounting to over 150,000 bytes of information, verified to the owner that the performance criteria had been met and that a high level of workmanship had been used in achieving the desired result.

REFERENCES

Federal Highway Administration, Dynamic Compaction for Highway Construction Vol. 1: Design and Construction Guidelines, July, 1986.

Melick-Tully and Associates, Inc., Soils and Foundation Report Proposed Home Depot Retail Facility, June 26, 1990.

Menard, L. and Broise, Y., "Theoretical and Practical Aspects of Dynamic Compaction", Geotechnique, March, 1975.

Slomski, Stephen, AMERICAN ELECTRIC POWER SERVICE CORPORATION, Personal Communications, 1993.

"Dynamic Compaction Used as a Winter Construction Expedient"
By
Lawrence F. Johnsen [1] and Christopher Tonzi [2]

Abstract:

Dynamic compaction allowed winter placement of weather sensitive fills for a 182,000 sf supermarket/warehouse in Danbury, Connecticut. As a result, the facility became operational on schedule and with significant budget savings.

In late Fall of 1989, grading for a supermarket/warehouse facility began. On-site fill materials consisted of silty sands, which are sensitive to moisture conditions during placement. With the onset of freezing temperatures the construction manager was seemingly faced with two alternatives: purchasing off-site granular fills, insensitive to moisture, or delaying the project several months. Since the owner found neither alternative acceptable the geotechnical engineer was asked to find another method.

The dynamic compaction method was chosen. It allowed the placement of up to 12 feet of silty sand fill during a record cold spell. The following July the fill was successfully densified with dynamic compaction using a 12.5 ton weight and drops of up to 60 feet. Test borings were taken to verify that satisfactory densification was achieved.

The placement of fill in subfreezing temperatures resulted in buried layers of frozen soil. The thawing time of the partially frozen fill was estimated by analysis, and later verified by test borings.

---

[1]Principal, Heller and Johnsen, Foot of Broad Street, Stratford, CT 06497, formerly, Associate Principal, GZA GeoEnvironmental, Inc.

[2]Geotechnical Engineer, GZA GeoEnvironmental, Inc., 27 Naek Road, Vernon, CT 06066.

Introduction

In the spring of 1989, a geotechnical investigation was undertaken for a proposed retail grocery building at a site located in a valley flood plain in Danbury, Connecticut. The proposed building dimensions were 320 feet by 570 feet, set at an elevation requiring approximately 12 feet of fill.

The eastern three-quarters of the site consisted of an undeveloped, grass covered field bordered by a small river. In the western quarter of the site, beyond the proposed building area, up to 26 feet of fill had been placed several years before to make the site accessible to an adjacent roadway.

The subsurface profile in the proposed building area consisted of 3 to 6 feet of silty granular fill overlying as much as 3 feet of organic clayey silt. These materials were underlain by varying thicknesses of silts and sands, and 65 feet of varved clay. The water table dropped 17 feet across the site with an average hydraulic gradient across the building area of 0.036.

Construction of the proposed retail structure on shallow footings and slab-on-grade required replacement of all existing fills and organic soils with controlled fill, as well as surcharging of the building pad to reduce post construction consolidation of the underlying varved clay.

The proposed construction was expected to start in early September, 1989. However, its start was delayed until October and fill placement in the building area did not start until November.

Construction Activities

In late November, 1989, the excavating contractor began removing existing fills and organic soils, and placing fill in the proposed building area. Since, in many areas, the removal of organic soils extended below the water table and the underlying naturally deposited soils were sensitive to disturbance, a 1.5 to 3 foot thick layer of shot rock was placed to provide a working mat above the water table.

It was intended to place on-site silty sands as controlled fill over the shot rock. Gradation of the fill materials is presented in Figure 1. However, compaction of the silty, moisture sensitive fills was not possible with the cold, wet weather encountered because of the late start of the project. Budget restraints made the importing of fill materials infeasible and the owner gave priority to meeting facility operating schedules. The geotechnical engineer then offered an alternative method of continuing with fill placement through the winter months and then, at a later date, compacting the fill with dynamic compaction.

The earthwork continued through December with fill placed in one to two foot thick lifts and compacted with either a vibratory drum roller or a vibratory sheeps foot roller. Density tests taken on the lifts showed compaction generally in the range of 75 to 90 % of the maximum dry density as determined by ASTM D 1557. During most of December, temperatures at the site remained below freezing causing overnight frost depths of several inches in the previous day's fill. In areas that did not receive fill the following day, frost depths reached two feet. Frozen soil was removed prior to placing fill when the frost depth exceeded six inches. The problems caused by the frozen fill layers could have been avoided by dumping the full depth of fill but there was concern that the resulting loose, saturated silty fills would not have sufficient strength to receive the drops of the dynamic compaction weight.

The project was temporarily shut down on December 29, 1989. By that time large portions of the building area had received 10 to 12 feet of silty sand fill. Upon resuming the project on February 7, 1990, temperatures were milder with average temperatures of 34.1°F for February and 40.3°F for March. The supply of on-site fills was exhausted by March and the contractor began importing a granular fill. Its gradation is presented in Figure 1. The imported sand and gravel fill was placed in lifts up to the final grade. The on-site fill had been placed to within two feet of final grade over most of the building area. In the north and west portions of the building area the thickness of on-site fill was less.

In April, a 4 to 5 foot high surcharge was placed over the entire building area in order to consolidate the underlying varved clay. In June, the consolidation was substantially complete and the surcharge removed. Settlements were monitored with reference plates placed at the base of the surcharge. Minor settlements may have occurred to the silty sand fill during the surcharge, but due to the location of the settlement reference plates that settlement can not be distinguished from the settlement of the varved clay.

In early July, test borings were performed over an evenly spaced grid throughout the area to receive dynamic compaction. The test borings verified that the layers of frozen fill had thawed and also provided a benchmark from which the effects of dynamic compaction could be evaluated.

Dynamic compaction began on July 13, 1990. All portions of the building area received three passes of a 12.5 ton weight dropping 60 feet. The initial and second pass drop points were located on overlapping twenty foot grids.

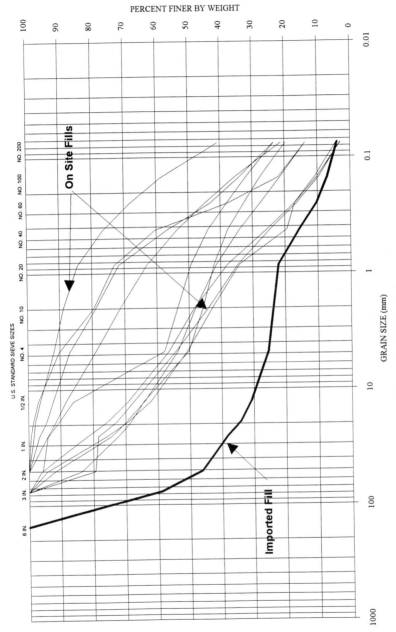

Figure 1. Fill Gradations

Third pass points were located at proposed column locations. The process of compacting the on-site fills during placement resulted in most of the subgrades having sufficient strength to receive the 60 foot drops of the weight. Crater depths were generally less than two feet. An additional series of test borings was taken following dynamic compaction to confirm compactive efforts. Standard penetration test results from borings taken before and after dynamic compaction are presented on Figures 2a and 2b to illustrate the effects of dynamic compaction.

One area of the silty sand subgrade along the west side of the work area remained saturated, probably due to the lateral flow of ground water across the site, and performed poorly under the 60 ft. drops. At this location, test boring B-8 showed that the upper six feet of silty sand had become looser, most likely due to the shearing action from adjacent drop points. During dynamic compaction in this area excess pore pressures were evidenced by large scale weaving of the surface and geysers of water entering craters. The craters were then leveled and up to three "ironing" passes consisting of 20 ft. drops of the 12.5 ton weight were performed in these areas. After two days of ironing passes, and sunny weather, the weaving stopped. A final series of test borings verified the improvement. Figure 3 shows both the detrimental effects of the 60 ft. drops and the beneficial effects of the 20 ft. drops in the area of the shallow, saturated silty sand fill. The dynamic compaction densified approximately 30,000 cy of fill in 27 working days.

Thawing of Fill

The decision to use dynamic compaction to compact fill after placement was made in early December after it became apparent that the silty sands with high water contents could not be adequately compacted in the predominantly 40°F temperatures. Normally December months in Danbury, Connecticut have average temperatures of 24.2°F and average high temperatures above freezing (Brumbach, 1965). However, shortly afterwards, Connecticut received record cold weather. For most of December the temperature at the site was continuously below freezing with an average temperature for the month of December of 19.1°F. Fill was placed and compacted in one to two foot thick lifts. Field measurements indicated that the fills were freezing to depths of up to 6 to 8 inches overnight. The manner in which the fill was placed resulted in a 10 to 12 foot thickness of alternating layers of frozen and unfrozen soil. It is estimated that approximately one-third of the fill placed in December froze during placement.

The utilization of dynamic compaction to densify fill placed during winter requires prediction and verification of the time required to thaw frozen fill.

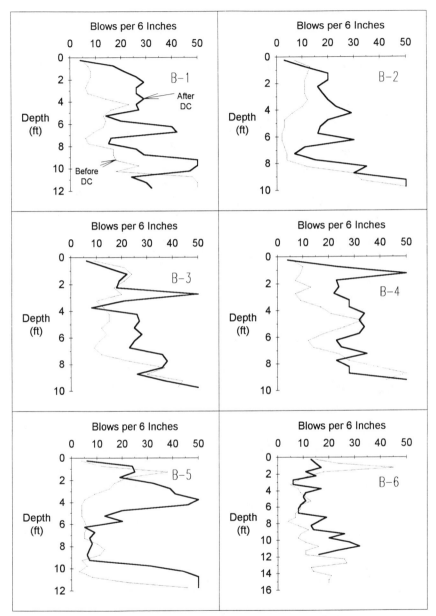

Figure 2a. Effects of DC 60 Foot Drops on SPT Blow Counts per 6 Inches

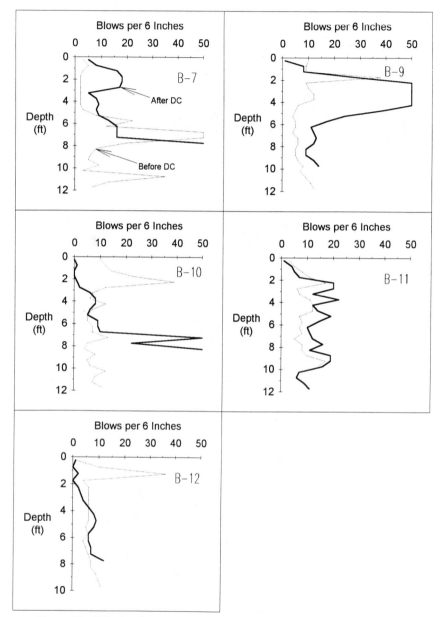

Figure 2b. Effects of DC 60 Foot Drops on SPT Blow Counts per 6 Inches

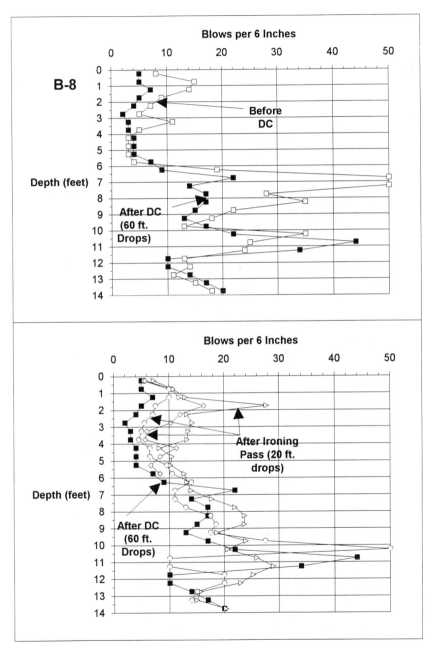

Figure 3. Effects of DC 60 and 20 Foot Drop Heights on SPT Blow Counts per 6 Inches

References 1 and 3 are listed to provide background on commonly used methods for predicting rates of freezing and thawing of soils.

The thaw analysis for this project was based on the basic heat flow equation: $Q = K \times i \times A$

Q is the latent heat of fusion and specific heat utilized per unit time

k is the coefficient of thermal conductivity

i is the thermal gradient of average surface temperature minus 32 degrees divided by the average distance the heat travels, which is one-half the total depth for a deposit for latent heat and one-third the depth for specific heat

A is unit area

Provided that the unfrozen layers are uniformly distributed, the effect of the unfrozen zones can be approximated by reducing the latent heat by a factor equal to the fraction of frozen soil. The quantity of specific heat was based on an initial average temperature of the fill of 32 degrees.

Assuming: 33 per cent of the fill was frozen; L (latent heat of fusion) = 1728 BTU/CF; X (depth) = 12 ft.; k = 1.5 BTU/FtHrF; c (specific heat) = .25; g (unit weight) = 120 pcf; and ΔT = (57.6-32)/2, the thawing time is $(570 \times 12 \times 12)/(2 \times 1.5) + (0.25 \times 120 \times 12.8 \times 12 \times 12)/(3 \times 1.5))$ = 39,936 degree hours, or 1664 degree days.

Records for the site indicate the following average monthly temperatures:

Month	Monthly Temperature	Degree Days	Cumulative Degree days
December	19.1	-400	
January	35.9	+121	+121
February	34.1	+59	+180
March	40.3	+257	+437
April	49.6	+528	+965
May	57.6	+794	+1759
June	70.1	+1143	+2902

Based on this analysis the fill had completely thawed by late May. Due to the required time for surcharging the varved clay, the dynamic compaction was not scheduled until July. Additionally, the method was conservative for this site as it ignores thawing from below. Considering that the fill was placed over a layer of shot rock and that the site had a significant lateral hydraulic gradient it would seem that the ground water within the shot rock would supply a significant amount of heat for thawing. Test borings taken in early July, prior to the start of dynamic compaction showed no evidence of frost.

On projects where the projected thaw time adversely affects the schedule, measures can be taken to minimize freezing and thermistors can be placed to verify the time at which thawing is complete. Freezing can be minimized by placing several lifts of fill in a limited area each day and placing insulation over unworked fill areas.

Equipment and Experience Required for Dynamic Compaction

The owner contracted with a crane contractor to provide the equipment and labor required to perform dynamic compaction under the direction of the geotechnical consultant. The contractor had no prior experience with dynamic compaction. The contractor supplied a Bucyrus Erie 71-B tracked crane.

Construction of a drop weight was required. Its design was based on guidelines provided in Reference 4. It was anticipated that the silty sands would be in a near saturated condition during the dynamic compaction operation. Therefore, a primary design concern was providing sufficient contact pressure to compact the soil, but limiting it to minimize the shearing of saturated silty sands. A contact pressure of 735 psf was chosen.

The resulting weight consisted of a three inch thick steel bottom plate and one inch thick steel side plates filled with reinforced concrete. It had dimensions of 4 ft. high, 6 ft. square with bevelled corners and a weight of 25,000 pounds. In general, the weight worked well with the silty subgrade. In one area an ironing pass of 20 ft. drops had to be performed to improve the silty sand subgrade prior to dropping the weight from 60 ft. heights. The construction manager provided a further improvement by attaching a blasting mat to the cable to limit damage from flying debris.

The crane operator practiced short drops of the 25,000 pound weight for about an hour before beginning production drops of 60 feet. The contractor experienced minor problems with cable wear, in particular, at the drum and the connection to the weight. The connection to the weight was changed from a single point with cable to a 4 point with chains to reduce cable wear. The cost of dynamic compaction including mobilization and 27 days of work was

$93,000, or $3.10 per cubic yard of fill. Since the completion of this project the contractor has successfully completed several other dynamic compaction projects.

Acknowledgements:

The authors wish to acknowledge the contributions of the project team including Stew Leonard, Jr., Tom Leonard and Frank Guthman of Stew Leonard's, Inc., Norwalk, Ct.; Richard Bergman of Richard Bergman Architects, New Canaan, Ct.; Arne Thune of Thune Associates, P.C., New Canaan, Ct., structural engineer; Rich Sonnichsen formerly of FGA Services, New Haven, Ct., site engineer; Al Bourett, John Navarro and Paul Taylor of The Facility Group, Inc., Smyrna, Ga., construction manager; Skip Gardella of Norwalk Marine Contractors, Inc., dynamic compaction contractor; and Pat Crowell, Tom Maier and Robert Heller formerly of GZA GeoEnvironmental, Inc., geotechnical engineer.

Additionally, the authors wish to thank Robert Heller of Heller and Johnsen for reviewing the paper and Sylvia Faustini for typing the manuscript.

Summary:

1. Dynamic compaction was found to be a viable winter construction expedient resulting in significant schedule and budget savings. Dynamic compaction allowed the placement of silty sand fills in below freezing weather with later densification by dynamic compaction achieving a suitable subgrade for support of foundations and slabs-on-grade.

2. Fine grained fills may require compaction during placement in order to have sufficient strength to densify under dynamic compaction treatment. This will result in buried frozen layers when freezing temperatures occur.

3. The practical use of dynamic compaction for densification of fills which freeze during placement requires an estimation of the thawing time of the fill. The use of the general equation for heat flow is considered reasonable when the layers of frozen soil are relatively uniformly distributed through the fill layer. Site conditions, such as the presence of a high ground water table and a high lateral hydraulic gradient may have significant effects.

4. Thaw times may be reduced by modifications to earthwork, such as, reducing the surface areas of lifts to reduce exposure time, insulating exposed lifts or removing excessive frost. Thaw times can be verified with test borings or thermistors.

5. The dynamic compaction method can be easily learned by crane contractors. This makes the method highly adaptable to remote sites or small projects where mobilization charges would otherwise make dynamic compaction uneconomical.

References:

1. H. Aldrich, Jr. and H. Paynter, "Special Report 104 - Depth of Frost Penetration in Non-Uniform Soil", U.S. Army Corps of Engineers, 1966.

2. J. Brumbach, "The Climate of Connecticut", State Geologic and Natural History Survey of Connecticut, 1965.

3. A. Jumikis, "Thermal Soil Mechanics", Rutgers University Press, 1966.

4. R. Lukas, "Dynamic Compaction for Highway Construction, Vol. I: Design and Construction Guidelines", FHWA/RD-86/133, 1986.

Dynamic Compaction of Saturated Silt
and Silty Sand - A Case History

Jean C. Dumas[1], Member ASCE, Nelson F. Beaton[2]
and Jean-François Morel[3]

Abstract

This case history describes the improvement of saturated fine grained alluvial soils using dynamic compaction. These conditions represent a borderline case for the application of dynamic compaction techniques. The site is situated on the north shore of the St-Lawrence River, between Montreal and Quebec City, in an area underlain by thick sequences of deltaic, marine and glacial deposits. The paper traces a broad outline of the methods used to enhance the ground improvement process and discusses the results of the accompanying monitoring and testing program.

Introduction

In 1992, a Price Club Wholesale Store/Warehouse was built in Trois-Rivieres-West, Quebec, on a site treated by dynamic compaction methods to allow construction to proceed with conventional spread footings and slab-on-grade. This ground improvement project is noteworthy for the considerable degree of difficulty it represented in view of the nature of the soils (loose saturated silts, sandy silts and silty sands) and the high ground water table condition. Special techniques, such as horizontal drainage and displacement methods, were used to enhance compaction and insure that the compactive energy could be applied at a rate compatible with the objectives of a tight construction schedule without, generating detrimentally high excess pore water pressures. Measurement of

---
[1]President, Geopac Inc, 1375 Joliot-Curie, Boucherville, Qc, Canada, J4B 7M4; [2]Vice-President, Geopac West Ltd, 615-8th Street, New Westminster, B.C. Canada, V3M 3S3, [3]Geotechnical Engineer, Geopac Inc., Boucherville.

the soil volume reduction and in-situ tests by the pressuremeter and the dilatometer methods were used to evaluate the initial compressibility of the natural soils, monitor the progress of the densification work and to optimize the compaction methods. The treatment was concluded by an extensive program of pressuremeter and dilatometer testing to verify the adequacy of work.

Site conditions

The site is situated at the confluence of the St. Maurice and St. Lawrence Rivers, on the north shore of the latter. It occupies what was formerly a low-lying area in which the ground water level occurred 100 mm below surface. It is underlain by thick strata of loose to compact fine-grained deltaic soils from the surface to a depth of 26 metres, slightly over consolidated marine clay to a depth of 31 metres and dense glacial moraine to an undetermined depth.

It was estimated that settlement generated by construction loads would, for the most part, occur in the deltaic soil deposit. The nature and looseness of these soils combined with a water table at 0.3 metres below original ground surface presented a distinct challenge for ground improvement.

The 26-metre thick deltaic deposit divides into two distinct soil layers. From the surface to a depth between 4 and 5 metres, the soils consist of loose to very loose silt having water contents ranging from 24 to 27 percent. Below, the deposit grades from loose sandy silt at the top to a more compact silty sand at depth.

The range of grain size distribution for these two soil layers and the range of Atterberg limits for the upper silt layer are shown in Figures 1 and 2 respectively.

Foundation selection

The 6.3-hectare Price Club property was raised with imported sand fill to an elevation about 2.5 metres above ground water level. The building occupies an area of 12,800 square metres and comprises a single-storey warehouse type structure with an attached two-storey office.

The average bearing stress applied over the building footprint area by the structural and live-loads amounts to 40 kPa. The maximum design columns loads are 800 kN. The maximum floor loads of 5,120 kN are applied over 8 m x 16 m rack assemblies. Footings were sized for an allowable bearing pressure of 200 kPa.

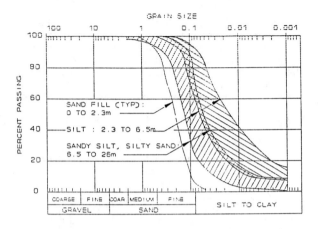

Figure 1. Range of Grain Size Distribution Curves

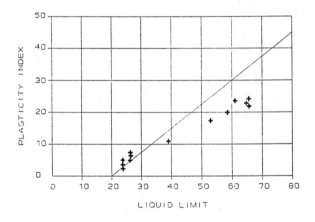

Figure 2. Plasticity of Upper Silt

Two foundation alternatives were considered: a fully structural solution based on piles, grade beams, and heavily reinforced floor slab; and a foundation on conventional footings and slabs-on-grade following ground improvement by dynamic compaction methods. The latter solution offered an estimated 1 to 1.2 million $ CAN economy in addition to reducing the construction schedule by months and was, therefore, preferred.

The compaction program

The objective of the treatment was to compact the sand fill required to raise the site to final grade and to densify the loose underlying native silt and sandy silt soils so as to limit total settlement to 25 mm for an allowable bearing capacity of 200 kPa under the footings and of 40 kPa under the slab. Treatment was carried out under the building footprint plus a 10-metre wide peripheral zone around the footprint.

A depth of influence of the order of 10 to 11 metres was considered necessary to attain those objectives and an energy per drop of 270 tonnes-metres was selected. It was estimated, on the basis of experience, that the application of 25 tonnes-metres of compactive energy per cubic metre of soil would achieve the necessary improvement.

The compaction plan called for the application of four high-energy phases on widely spaced points followed by a final low-energy full site coverage, or ironing phase. To avoid the build-up of high excess pore water pressures in advance of the compaction front, the first phase compaction points were set on a square grid 14 metres on center and each successive phase was carried out on a grid set at the center of the preceding phases compaction points.

Densification to the full depth of influence was to be achieved with the first two phases. The third and fourth phases were to densify the loose upper soils in between, and the ironing phase was designed to tighten the top 2 metres.

To enhance compaction and accelerate the dissipation of excess pore-water pressure, particularly in the upper silt, phases 1,2 and 3 were to be carried out using a sand displacement technique leading to the formation of sand columns 3 to 3.5 metres in diameter and 5 to 6 metres deep (Dumas, Beaton, Morel, 1993).

The compaction plan also provided for the installation of a horizontal drainage system designed to intercept the water expelled as a result the soil volume reduction. A series of pneumatic piezometers were installed to evaluate the pore-pressure response and to determine the time-lag required between successive phases of treatment.

Execution of the ground improvement work

Site preparation began in early March 1992 with the removal of snow and topsoil. The compaction area was raised to final grade with SW fill material, leaving the surrounding area at original grade elevation. A horizontal drainage system was then installed under the compaction footprint. It consisted of 100 mm diameter plastic perforated pipes installed in parallel rows 7 metres apart which discharged into a main collector and from there into a transfer well. As spring break-up was imminent, a time of the year when the site used to be submerged under 300 to 600 mm of water, a peripheral ditch was dug around the built-up area and connected to the transfer well through a discharge separate from that of the horizontal drain system so as to allow the discharge from horizontal system to be monitored.

Preparatory work also included the installation of pneumatic piezometers and the performance of pre-treatment in-situ tests. A total of ten piezometers were placed at depths ranging between 4.7 and 9.1 metres at 5 locations. Twelve pre-compaction boreholes were performed, four by the Menard pressuremeter method and eight by the Marchetti dilatometer method.

Compaction of phases 1,2 and 3 compaction points was done using the sand displacement method mentioned earlier. Compaction of phase 4 and the ironing phase was done following standard dynamic compaction procedures. Field tests were carried out at various intervals to determine the rate at which energy could be applied without causing excessive pore pressure and heave. The result was a procedure which consisted of expanding the sand columns in three to five increments leaving sufficient time between each increment for dissipation of excess pore pressures. The piezometer record shown in Fig. 3 gives an account of the progress of the operation and the success achieved in controlling excess pore-pressures.

The total energy applied amounted to 4 million tonne-metres, representing an average energy of 27 tonnes-metres per cubic metre of soil treated. A total of 10,800 cubic metres of sand went into the formation of sand columns. It was determined that the phase 1 sand columns reached a average depth of 5.5 metres. The depth of phases 2 and 3 columns was not determined, but assuming it was proportional to the energy applied on those compaction points, it is estimated that it averaged 5.5 at phase 2 points and 3.3 metres at phase 3 points.

The horizontal drainage system began discharging water a few hours following the start of the treatment and remained operational over a period of 25 days , until it

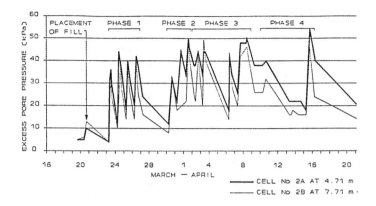

Figure 3. Record of Excess Pore Water Pressure

was collapsed in the fourth compaction phase. The measured discharge at compaction time varied between 180 and 280 litres per minute. Allowing for the reduction in discharge at night and during week-ends, it is conservatively estimated that the volume of water discharged by the system as a result of compaction phases 1, 2 and 3 was of the order of 7,000 to 9,000 cubic metres. Excavation performed shortly after compaction indicated that the upflow of water had ceased, indicating that the flow was a direct result of volume reduction during compaction.

DMT testing was used extensively to monitor the progress of the treatment and verify final results. A total of twenty four DMT boreholes were done during and after compaction. Pressuremeter tests were also performed in 10 boreholes using the standard three-cells Menard probe and a GAM read-out unit. The AW probe was inserted into the ground by means of a driven slotted tube.

The site was delivered to the general contractor on May 6 1992, exactly two months after the first day of mobilization on March 7.

Evaluation of the improvement achieved

The effectiveness of the Trois-Rivieres ground improvement work can be judged from the amount of soil volume reduction achieved and from the comparison of the before and after PMT and DMT test results.

The soil volume reduction achieved by the treatment, as determined from the volume of fill that went into the formation of sand columns corrected for the change in elevation of the work platform, and from the depth of improvement indicated by testing, amounted to approximately 7.2 percent. This is well above the 5 to 6 percent historical average for fine-grained alluvial soils.

The comparison of before and after PMT and DMT test results confirms the efficiency of the treatment and, as shown in Fig. 4 and 5, provide a detailed profile of the degree and depth of improvement. The diagrams in Fig. 6 shows statistically the shift of PMT moduli and pressure limits towards higher values.

Predicted foundation performance

Post-treatment PMT and DMT test data was used to assess foundation performance in terms of bearing capacity and settlement. Predicted total settlement under isolated and strip footings varied from a low of 5 mm to a high of 10 mm for a design bearing stress of 200 kPa. Predicted total settlement under the slab varied from a low of 6 mm to a high of 23 mm. Allowable bearing capacities of between 250 to 500 kPa were available following treatment.

Calculations were made according to the Menard method in the case of the PMT data (Centre d'Etudes Geotechniques, 1964) and methods proposed by Schmertmann (Schmertmann, 1988) were used for the DMT data: the q_D method for

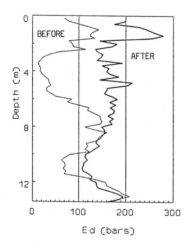 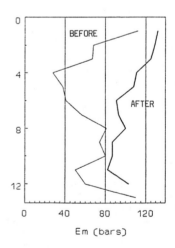

Figure 4. Comparison of the Before and After Average Marchetti DMT Moduli, Ed, and Menard PMT Moduli, Em

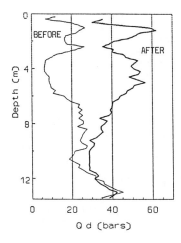 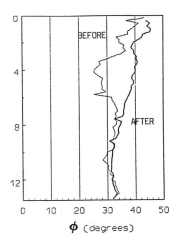

Figure 5. The Diagram on the Left Shows the Increase of DMT Resistance to Penetration, q_D, and that on the Right, the Corresponding Increase in the Angle of Internal Friction as Calculated from q_D

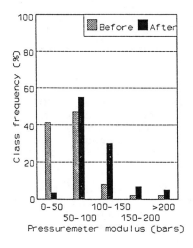 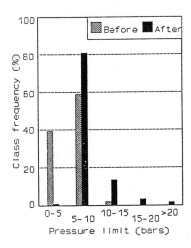

Figure 6. Statistical presentation by the frequency distribution method of the shift of PMT moduli and pressure limits towards higher values

the calculation of the allowable bearing capacity and the Janbu method for the calculation of settlement. The results of these calculations are summarized in Tables 1 and 2. In spite of some differences, these various calculations provided ample assurance that the objectives of the treatment had been achieved.

Table I   Summary of Bearing Capacity Calculations

Method	Column Footings, kPa		Strip Footings, kPa	
	Range	Average	Range	Average
PMT	406 to 613	508	320 to 707	435
DMT	249 to 637	428	236 to 476	351

Table II   Summary of Settlement Calculations

Method	Col. Footings mm		Strip Footings mm		Slab mm	
	Range	Av.	Range	Av.	Range	Av.
PMT	6 to 9	8	5 to 10	8	6 to 8	7
DMT	2 to 8	3	1 to 9	3	8 to 23	14

Calculations were also made from pre-treatment DMT test data to evaluate what the settlement would have been without ground improvement. These results showed that the settlement under columns would have ranged from 21 to 198 mm for an average value of 108 mm, over 30 times the average post-treatment results.

According to Schmertmann, the q_D obtained from the dilatometer is substituted for the q_c in the CPT-based bearing capacity methods thus providing an alternate method for using the DMT data to evaluate bearing capacity.

Conclusion

The extensive program of PMT and DMT testing performed for this project provides well-documented evidence of the efficiency of the Dynamic Compaction method for improving saturated silts and sandy silts in the presence of high ground water table.

The bearing and settlement characteristics achieved on this site are considerably greater than those normally anticipated for these types of soils and water table conditions. This superior response can only be attributed to the special methods of compaction and drainage implemented on this site.

The work was carried out on the basis of a performance specification, for a firm lump sum price of 398,700 $CAN. The price includes the drainage system, instrumentation, monitoring and all the PMT and DMT verification testing. The project was completed in two months, satisfying the time allowed for in the construction schedule. The fill material was not included.

References

CENTRE D'ETUDES GEOTECHNIQUES, 1967
Rules for the Use of Pressuremeter Techniques and Processing the Results Obtained for the Calculation of Foundations, Notice D/60/67.

DUMAS, J.C., BEATON, NELSON F. and MOREL, JEAN-FRANCOIS, 1993
Dynamic Compaction Using Select Fill Displacement Methods, Proceedings:Third International Conference on Case Histories in Geotechnical Engineering, St-Louis Missouri, Paper No. 7.31

SCHMERTMANN, JOHN H, 1988
Guidelines for using the CPT, CPTU and Marchetti DMT for Geotechnical Design, Vol IV - DMT Design Methods and Examples, Report No FHWA-PA-024+84-24.

PROSPECTS OF VACUUM-ASSISTED CONSOLIDATION FOR GROUND IMPROVEMENT OF COASTAL AND OFFSHORE FILLS

S. Thevanayagam[1], Member, ASCE, E. Kavazanjian, Jr.[2], Member, ASCE, A. Jacob[3], Member, ASCE, and, I. Juran[4], Member, ASCE

ABSTRACT: Vacuum-assisted consolidation can be an effective alternative for conventional surcharging to improve fine-grained soils. Cost and implementation difficulties were the major drawbacks for its effective use in engineering practice. Recent technological advances in the application of geosynthetic materials have now made on-land vacuum consolidation an attractive and cost-effective alternative to conventional mechanical surcharging for improvement of compressible soils. Experience from recent field applications of on-land vacuum consolidation at the Port of Los Angeles, Tianjin Harbor-China, Japan, and France indicates that the technology holds technical merit and is cost-effective compared to other conventional techniques. Potential uses of vacuum technology currently under consideration by the Port of Los Angeles include improvement of deposits of soft soils in the existing hydraulic fills, strengthening weak foundation soils on the sea floor below the fills or adjacent to waterfront retaining facilities, and accelerated consolidation of fine-grained hydraulic fills during construction. Application of vacuum-assisted consolidation in the latter cases potentially offers significant advantages.

INTRODUCTION

Dredged fill placement techniques are increasingly being used in several ports, harbors and coastal areas to reclaim hundreds of acres of land for facility expansion. Often, such reclamation projects encounter weak foundation soils (e.g. Chek Lap Kok Airport, Koutsoftas and Cheung 1994, New York LaGuardia Airport Runway Extension) and/or fine-grained source material for the fill itself (e.g. Port of Los Angeles 1987, Jacob et al 1994, Tokyo Airport, Tanaka 1994). Such projects

---

[1]Asst. Prof., Dept. of Civil Engrg., Polytechnic Univ., Brooklyn, NY 11201

[2]Assoc., Geosyntec Consultants Inc., 16541 Gothard St., Huntington Beach, CA 92647

[3]Civil Engrg. Associate, Port of Los Angeles, San Pedro, CA 90733

[4]President, GIT Consultants Inc., 707 Westchester Ave., White Plains, NY 10604; Prof. and Head, Dept. of Civil Engrg., Polytechnic Univ., Brooklyn, NY 11201.

traditionally rely on conventional dredge material placement and surcharged stabilization techniques for the construction of fills.

In the case of fine-grained dredged fills, experience indicates that a waiting period of two to three years or more of desiccation and consolidation due to self-weight is required before construction equipment can access fills for installation of vertical drains and surcharging. Once the site is 'walkable', stage construction of the preload is generally often required due to the low initial shear strength of the hydraulic fill, even with vertical drain installation. Due to cost of handling large quantities of surcharge fill material and the need for eventual disposal, often, only a portion of the site is preloaded at one time using a limited amount of surcharge fill. Once consolidation of one phase is complete, the surcharge is moved to another location for the next phase of surcharging. It is not unusual for a 200-acre site to require a decade or more of post-landfilling development before it can be used. In addition to the direct costs of surcharging and the loss of revenue and/or interests costs on initial capital investment associated with a long fill development period, the indirect costs of ground improvement by surcharge fill is steadily increasing, particularly in urban areas. Indirect costs include environmental costs associated with emissions from earth-moving equipment, dust control, and traffic impacts generated by the need to place, remove, and dispose of surcharge fill materials. These rising direct and indirect costs have created a demand for new technologies to expedite the self weight consolidation and stabilization of fine-grained hydraulic fills.

In the case of stabilization of soft foundation soils beneath fills, retaining structures and breakwaters, often times, staged construction may be required to maintain adequate stability, even when prefabricated drains are used to accelerate dissipation of excess pore pressures and induce concurrent strength gain. Surcharging and preloading of soil at and beyond toe of these structures may not be technically or economically feasible. Furthermore, high costs can be associated with the delays in facility development due to the need for staged construction and surcharging. Techniques for accelerating and enhancing the strength gain in these situations can also provide economic benefits.

Another problem often faced in coastal/port facility development is the need for large storage capacity for disposal of dredge material from deepening of navigation channels. Similar problems are encountered in the case of disposal of waste slurries/tailings. Stability and design life of such storage facilities can be significantly improved by reducing the water contents of these suspension-like materials.

This paper presents case histories of on-land vacuum consolidation to accelerate consolidation of soft soils. Potential future coastal applications are presented. Technological challenges and solutions are discussed. Research needs are identified for full scale practical implementation of this technology for the latter applications.

BACKGROUND

The concept of vacuum consolidation was first proposed by Kjellman (1952) in the early 1950's. Instead of increasing the effective stress in the soil mass by increasing the total overburden stress as in conventional mechanical surcharging, vacuum-assisted consolidation relies on increasing effective stresses by decreasing the pore pressure while maintaining the existing total overburden stress. In the 1970s, the Corps of Engineers conceptually investigated the feasibility of vacuum consolidation of dredge disposal areas to increase storage capacity (Johnson et al 1977). Isolated studies of vacuum induced or assisted consolidation continued for the next two decades (Halton et al. 1965, Holtz and Wager 1975). However, except for specialized applications like landslide stabilization (Artunian 1976), vacuum consolidation was not seriously investigated as an alternative to conventional surcharging until recently due to the low cost of conventional surcharge technology, the lack of technology for sealing the area over which the vacuum is applied, and other difficulties involved with applying and maintaining the vacuum. The advent of technology for installation of prefabricated systems and for sealing landfills with impervious membranes for landfill gas extraction systems have now made *on-land vacuum-assisted consolidation* (Fig.1) an economically viable method as a replacement for or supplement to surcharge fill.

Vacuum technology combined with prefabricated drainage systems can also be an attractive means to expedite self-weight consolidation of new engineered fills, and waste slurries / dredged fill in disposal areas (e.g. Figs.2-3).

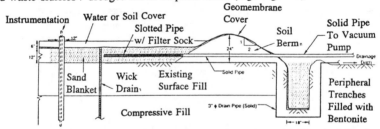

FIG. 1. Schematic On-Land Vacuum Consolidation with Vertical Drains

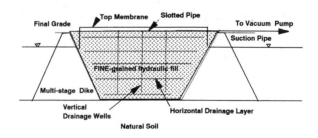

FIG. 2. On-Land Vacuum Consolidation with Vertical and Horizontal Drains

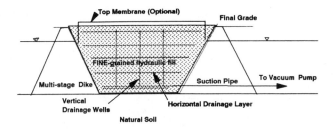

FIG. 3. Underwater Vacuum Consolidation with Vertical and Horizontal Drains

DESIGN ASPECTS

There are many technical and operational factors which play important roles in vacuum consolidation. The success of a vacuum consolidation system depends upon implementation of a combination of technological know-how and design details. The potential for successful application of this technology is very much dictated by site-specific soil types, soil stratigraphy, and groundwater conditions. The technology is very versatile. The vacuum can be applied from the top of the ground surface (*on-land application, e.g. Figs.1-2*) or from below the water level beneath the fine-grained deposit (*underwater application, e.g. Fig.3*). The equivalent preload attained by these cases are significantly different (as demonstrated subsequently in Fig.11). Therefore implementation strategies are likely to differ significantly from place to place and depends on the specific project needs. The mechanism of consolidation and the rate of settlement in vacuum-assisted consolidation are the same as for conventional surcharging. Hence, geotechnical consolidation design analyses used to evaluate drain spacing, settlement rate, and strength gain for preload fills are equally applicable to vacuum consolidation. Unlike surcharging, which may cause lateral spreading of the soft soil and pose stability concerns, vacuum technology does not impose any external load and therefore does not induce any additional destabilizing load on the soil.

ON-LAND VACUUM CONSOLIDATION

Typical System Configuration: A typical on-land system consists of an air-tight membrane over the ground surface and anchored along the periphery of the site (Fig.1). Slotted collection pipes are embedded into a sand blanket beneath the membrane. The outlet is a solid pipe and connected to a vacuum pump and discharge system. Typically, for an existing soft ground site, vertical wick drains are installed beneath the sand blanket up to a depth of about 1 m above the bottom of the soft soil deposit. Usually application on an existing site for on-land application requires that the site be 'walkable' so that the equipment can access the site for installation of wick drains and placement of membrane seal. For new hydraulic fills an initial waiting period may be required. An engineered fill with continuous horizontal pervious layers or prefabricated drainage sheets, if properly placed, could conceivably be vacuum-consolidated with only a limited number of large diameter vertical-drains (Fig.2). Unlike in conventional sand-clay layered fill construction, the high permeability horizontal layers in such a system should be confined within the fine-grained fill area to reduce potential leakage of water through them during vacuum application.

Cavitation of water at negative 1 atmosphere limits on-land vacuum-consolidation to an effective surcharge pressure of about 100 kPa, equivalent to about 6 m of surcharge. Practical problems in maintaining the efficiency of a vacuum system may reduce its effectiveness in the field. A system with an efficiency of 75 percent results in only 4.5 m of equivalent surcharge.

Primary considerations governing the effectiveness and economy of an on-land vacuum consolidation scheme include: (a) integrity of the membrane itself and the seal between the edges of the membrane and the ground, (b) interaction between wick drains and subsurface layers of high hydraulic conductivity which extend beyond the bounds of the membrane, and (c) depth to groundwater (e.g. Cognon et al 1994, Jacob et al 1994). Fig.4 shows anticipated pore pressures and effective stresses due to on-land vacuum consolidation with vertical drains.

Because a vacuum-assisted consolidation system is not meant to be a permanent system, ultraviolet degradation of the membrane is not typically an issue. Leakage through defects in the membrane or the seal between the membrane and the ground and drains extending through laterally continuous granular layers of high hydraulic conductivity within or below the cohesive soil deposit being consolidated can inhibit efficiency, reducing equivalent surcharge height. If the pervious layers are discontinuous and/or do not extend beyond the edge of the membrane, horizontal layers of pervious soil can be beneficial, increasing the rate of consolidation by inducing vertical as well as radial flow and thus increasing the allowable vertical drain spacing and shortening the duration of vacuum system operation. This, in turn, will reduce the cost of operation by reducing pumping costs and/or drain installation costs.

On-Land Field Trial Applications: Recent field trials in China (Choa 1989, Liu, undated), the USA (Jacob et al 1994) and France (Cognon 1991, Cognon et al 1994) have verified the effectiveness of on-land vacuum-assisted consolidation in

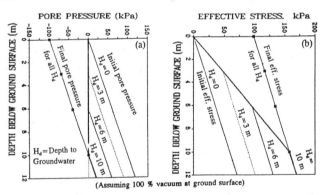

FIG.4. Pore Pressure and Effective Stress Profiles due to On-Land Vacuum Consolidation

conjunction with prefabricated vertical band drains. Field trials in Japan have proven effectiveness of vacuum consolidation of very soft reclaimed lands /waste slurries.

Port of Los Angeles - Pier 300 Site

Test Section: The test section at Pier 300 site consisted of a 30x30 m² area consisting of prefabricated band wick drains (100 mm x 3 mm) at a spacing of 1.5 m capped with a 0.05 cm thick (20mil) PVC membrane, a perimeter anchor trench for membrane seal and a vacuum pump. The membrane was covered with about 0.3 m deep water. In vacuum consolidation, the cost of pump operation can be large compared to wick drains. Therefore, to reduce pumping costs a wick drain spacing of 1.5 m was used to limit the consolidation duration to less than 2 months.

Fig.5 shows the typical soil profile across the vacuum test pad area. The groundwater elevation was about 3.3 m (11 ft) below the ground level. The fill consisted of intermittent thin seams of sandy silt or clayey silt within relatively thick clay layers with large variations in consistency and characteristics. Wick drain installation was typically terminated at about 0.6 m above the bottom sand. Due to the variability of the depth to this sand deposit, termination depth was determined on a location-by-location basis based on the penetration resistance felt by the operator of the wick drain installation equipment. Initially the contractor chose to install a very shallow trench for the membrane seal around the test pad. Due to significant leakage of air, then he chose to install about a 3-m deep and 2-inch thick in situ soil-bentonite slurry mix. Subsequently a cut-and-cover membrane-soil anchor trench system was constructed up to a depth of about 3.3 m. The membrane was covered with water retained by a small (0.3 m high) perimeter berm.

Observations: After several initial trials in which the membrane seal system was modified from a very shallow trench to a cut-and-cover system and the pump size was increased from 0.5 to 10 horse power, a vacuum pressure of about -80 kPa was attained beneath the membrane. However, this initial pressure lasted only for about a day and with time the vacuum pressure dropped to about -65 kPa and was sustained thereafter to the end of the test. Once the vacuum was established using the 10 horse power pump, it was only necessary to install a 1/2 horse power pump to sustain the vacuum. Fig.6 shows the profiles of measured pore pressures and interpreted effective stresses at different stages of perimeter membrane seal enhancement and at the end of the test. For comparison, profiles corresponding to the initial state of stress prior to vacuum application and the theoretical maximum corresponding to 100 percent efficiency in vacuum consolidation are also presented.

Negligible changes in pore pressures were observed during the operation with the initial very shallow membrane anchor trench. Significant changes in pore pressures were observed when a thin slurry cut-off wall was installed around the perimeter of the test pad. However, the pore pressures began to increase after a few days of operation. This appears to have been caused by increase in leakage of air as the thin slurry wall began to dry and desiccate with time. With the cut-and-cover membrane seal, however, a significant reduction in pore pressures were observed. This was sustained without significant losses. These observations indicate the critical nature of the membrane seal in achieving high efficiency.

96 DEEP SOIL IMPROVEMENT

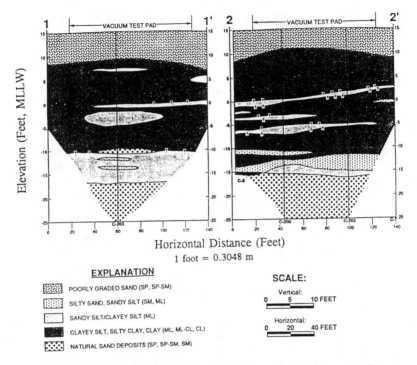

FIG. 5. Typical Soil Profile at Test Area - Port of Los Angeles

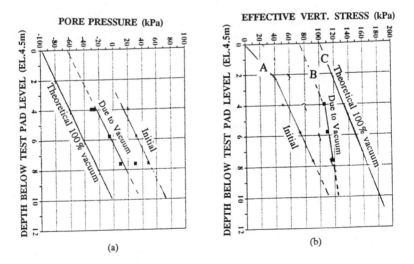

FIG. 6. Measured Pore Pressures and Effective Stresses - Port of Los Angels

Vacuum Consolidation Efficiency: The overall average efficiency (average increase in effective stress divided by atmospheric pressure of 100 kPa), was estimated between 40 to 50 percent, equivalent to about 3 m of surcharge fill. The efficiency varied with depth between about 65 percent at the top of the cohesive hydraulic fill to less than 30 percent at the bottom of the cohesive layer. Examination of the effective stress profiles at the end of test in Fig.6b and the subsurface stratigraphy (Fig.5) indicates that the losses in efficiency at greater depths depth could be due to: (i) seepage through the pervious lenses within the cohesive fill, and/or (ii) seepage from the bottom marine sand layer into the wick drains. Size effects of the test area may have also impacted efficiency. In a small test area, boundary leakage has a relatively large influence. For a larger area, the losses across the boundary become a proportionally smaller percentage of the total volume of fluid extracted during consolidation and the relative efficiency of the system should increase. Alternatively, a deeper perimeter cut-off system (such as recent developments in plastic sheet piles) could also reduce the loss due to perimeter leakage through pervious lenses. In such a case, the presence of pervious layers would become beneficial in accelerating the consolidation process.

Effect of Layered Soil Stratigraphy: The above observations indicate the influence of soil stratigraphy on vacuum consolidation. Without a vertical cut-off system vacuum consolidation can sometimes be ineffective. Due to variations in the dredge material source, hydraulic fill discharge locations and placement techniques significant amount segregation of coarse and fine grained materials can occur. As a result certain areas of pier 300 site consisted of rather thick granular layers sandwiched between clay layers. The field trials at Pier-300 consisted of another test section, similar to the test section described above, constructed at such a layered zone as well. Initial vacuum pumping operations did not yield any significant changes in pore pressures beneath the membrane or in the subsurface.

French Experience

Several soft ground sites consisting of different soil types including peat has been vacuum consolidated in France. A detailed summary of French experience is presented elsewhere (Cognon 1991, Cognon et al 1994). The sites include highways, airports, oil tank farms etc. The depth of improvement at these sites ranged up to 12 m. Ground settlements observed due to vacuum consolidation were about 10-20% of the thickness of the soft layers, a typical range encountered due to conventional surcharging. Ground improvement work at some of the sites in France included placement of additional surcharge on top of the air-tight membrane to obtain a larger equivalent preload. These field trials have proven the feasibility of vacuum consolidation combined with conventional surcharge.

Tianjin Harbor Site - China

The Tianjin Harbor site consisted of about 4 m thick hydraulic fill (thinly laminated silty clay/clayey silt) underlain by 15 m of silty clay, 3m of clayey loam, and another 3 m of sandy loam, respectively. The site overlies a fine sand deposit. The groundwater level was very close to the ground surface at this site. Vertical band drains (100mm x 4 mm) were installed at a spacing of 1.3 m up to a depth of about 20 m into the clayey loam above the sandy loam. Pore pressure measurements at a test section are shown in Fig.7 (Choa 1989). The results are similar to that observed in Pier 300 site. A vacuum consolidation efficiency of about 70 percent is observed near the ground surface. At large depths, however, the efficiency was only about 50 percent or less. The loss of efficiency with depth appears to be related to the presence of relatively permeable layers at large depths.

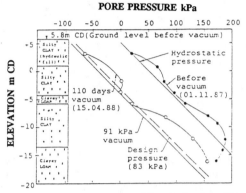

FIG. 7. Measured Pore Pressures - Tianjin harbor Site (Choa 1989)

Japanese Experience - Consolidation of Very Soft Reclaimed Land
Recent field trials in Japan have verified the effectiveness of vacuum assisted consolidation to reduce the water content of very soft ground sites reclaimed by dredge fill material or waste materials such as soda ash (Shinsa et al, undated, in Japanese; Shinsha et al 1991). The primary purpose of consolidation in these field trials was to assess the potential for using vacuum consolidation to reduce the water content of slurry-like material disposed at temporary storage areas and subsequent transport of the material for off-site disposal. The vacuum-consolidation implementation schemes in these cases are significantly different from the above typical on-land applications. Wick drains (100mm x 3mm or 150mm x 12mm) were placed horizontally (or vertically) at a spacing of 0.7-1.5 m within the dredged fill from a lightweight barge after fill placement (e.g. Fig.10 for horizontal placement). Each of the wick drains is connected to a hose from the vacuum pump at the ground level. Unlike in the case of the on-land applications discussed earlier, the air-tight membrane is absent in this system. The soft soil above the top wick drain acts as an air-tight seal (In the case of vertical drains each of the wick drains is terminated about 1 m below the top of the soft ground; a hose is attached to the wick drain at that point and connected to the vacuum pump; the corresponding figure is not shown; Both methods of wick installation provided similar ground improvements). Fig. 11 shows the water content distribution prior to and after application of the vacuum consolidation at the soda-ash site. A significant reduction in water content from about 300-400 percent to less than 200-300 percent was achieved. In general, ground settlements up to about 40-50 percent of the thickness of the soft deposits

could be achieved by this method and therefore can be used to significantly increase the storage capacity of disposal sites. After new material is placed it could be consolidated in a similar manner.

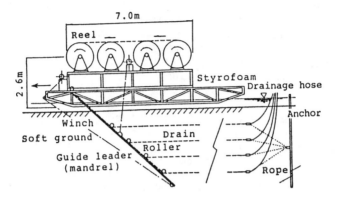

FIG. 8. Schematic Horizontal Wick Drain Installation (Shinsa et al 1991)

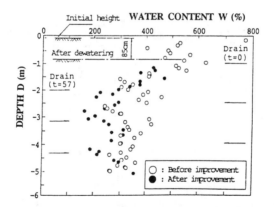

FIG. 9. Water Contents Profile at Soda-Ash Site (Shinsa et al 1991)

On-Land Application combined with Dewatering

The effectiveness of the on-land system could be further enhanced and the consolidation process could be further expedited if the vacuum pumps are periodically stopped and the vertical drains are dewatered (Fig.10). After dewatering the drains, the vacuum pumping is resumed again. During this vacuum pumping, full vacuum would be felt for a larger depth up to the water level in the drain.

Field trials during ground improvement work at the international airport site in Malaysia (Woo et al 1989) have partially demonstrated effectiveness of such a system using sand drains. The site consisted of soft clay up to a depth of 15 m underlain by 10 m of stiff clay and dense sand at the bottom. Sand drains (26 cm diameter) with small tubings (Fig.10) were installed within the soft clay deposit up to 14.5 m depth at a spacing of 2 m x 1.75 m. The site was sealed with impermeable membrane in a similar manner shown in Fig.1. Comparison of trends in the measured settlement data and rates observed in those trials indicated promise for such an approach. Conceptually such a system can be adopted for wick drains as well except it may require frequent alternating between dewatering and vacuum pumping due to small volume of wick drains. Wick drains could be modified slightly using a slotted tubing instead of the plastic core in the band drains to increase the capacity.

The system shown in Fig.2 (discussed earlier) with large diameter wells could be operated in a similar manner described in Fig.10. Depending on the dewatering operational details it could yield equivalent preload as high as an underwater vacuum system (described later in the next section, Fig.3).

As illustrated in each of the above cases, essentially the same concept of on-land vacuum consolidation has been applied in different forms, depending on the soil characteristics, site stratigraphy, required equivalent preload level and time constraints. In some of the above cases even when conventional surcharging is practically impossible vacuum-consolidation provides an effective solution.

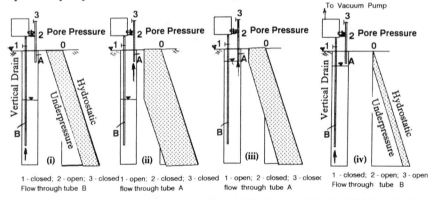

FIG. 10. Schematic On-Land Vacuum Consolidation combined with Dewatering

UNDER-WATER VACUUM CONSOLIDATION

General Principle: By applying the vacuum below the water table so that it is used in combination with dewatering, the equivalent preload can be increased significantly (Figs.3 and 11). Such a system employs the net reduction in pore pressure due to vacuum application and dewatering to increase the effective stress in the soil. This type of system appears particularly suitable to dewater dredge spoils and to improve new hydraulic fills. Horizontal drainage layers can be engineered within the fill during construction by placing sand, prefabricated drainage sheets or wick drains as the fill is placed (In earthquake prone areas use of prefabricated synthetic drainage layers may be more prudent than sand drainage layers, due to high potential for liquefaction of confined sand layers and subsequent instability). The drains should be confined within the soils to be improved to reduce leakage from outside the fill. Large diameter vertical drainage wells or horizontal headers at the ends of the drains may be used to drain the water from the horizontal drains.

Equivalent Preload: Fig.11 shows typical initial and final pore pressure and effective stress profiles that could be attained due to underwater vacuum consolidation assuming 100 percent (A) and 50 percent (B) efficiency. Ground improvement due to self weight consolidation alone is also shown in Fig.11. For comparison, Fig.11 also shows the maximum effective stress profiles corresponding to an on-land system operating at 50 and 100 percent efficiency. The equivalent preload for an underwater system is significantly higher than an on-land system. Even if the system is operated solely as a dewatering system without vacuum assistance, a method sometimes referred to as underpressure consolidation (Fig.12) a significant increase in effective stress can still be attained due to the reduction in pore pressure and increase in effective stress (C in Fig.11).

Efficiency: Primary considerations governing the effectiveness and economy of under-water vacuum or underpressure consolidation scheme include: (a) integrity of encapsulation of the horizontal drainage layers within the fine-grained soil, and (b) the method of applying the vacuum/dewatering system to the soil deposit. Unlike the on-land applications, the depth to groundwater does not adversely affect the equivalent preload that can be attained in such a system (A, B in Fig.11).

Leakage through the horizontal drains or top and/or bottom of the improved section can inhibit the system efficiency, reducing equivalent surcharge height. One solution to this problem is a total encapsulation of the entire fill (Fig.3). An encapsulated system requires careful placement of horizontal drains such that they are confined within the fine-grained soil and possibly placement of membranes below the hydraulic fill prior to fill construction. Unlike in the on-land system, construction of joints between adjacent panels of membrane may be not be a critical factor. The system should be self-sealing to some extent due to confinement of the overlaps of membranes due to the weight of the fill above the membrane and the vacuum that will tend to close any gaps along the panel overlaps. If necessary, a hydrophilic band could be attached along the edges of the membrane panels so that they would expand when wetted and further reduce the potential for leakage. To obtain maximum benefit, the design details with regard to the thickness of encapsulating soil along the boundaries, membrane seam integrity, spacing and extent of horizontal drains, vertical extraction wells or headers, and vacuum application technique should be carefully evaluated prior to construction of the fill.

Field Trials: A full-scale pilot study has not yet been done on application of underwater vacuum consolidation for improvement of hydraulic fills. The experience from Japan discussed earlier is essentially an on-land system except the use of horizontal band drains. To the authors' knowledge, let alone vacuum consolidation, prefabricated horizontal drains alone as a substitute for sand layers have not yet been used in hydraulic fills.

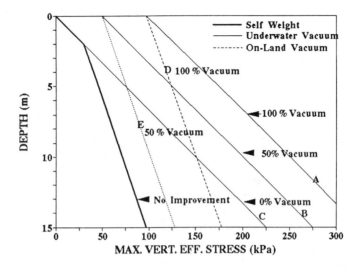

FIG. 11. Effective Stress Profiles Due to Vacuum Consolidation

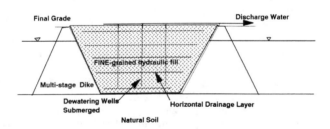

FIG. 12. Ground Improvement by Dewatering

Assuming a 15 m high hydraulic fill and a 2 m of depth to groundwater, and assuming that 6 m of equivalent surcharge is required for ground improvement, Table 1 summarizes the equivalent preload that could be attained, the typical waiting period for the start of fill improvement, and the typical duration required for improvement for underpressure consolidation, and conventional surcharging. Depending on site conditions, material cost, and design details, the total and relative cost will vary from project to project. However, the advantages of shorter overall duration of the fill construction and ground improvement and therefore early revenue is present for all vacuum and underpressure consolidation applications.

Table 1: Comparison of Equivalent Preload, and Duration

Preload Type	Drainage	Surcharge/ Equivalent Preload (m)	Initial Waiting (yrs)	Preloading (yrs)	Total (yrs)	Figure No.
Multi-Stage Surcharge	Vert. Wicks	6	3-5	2-3	5-8	--
	Horiz. Drains @	6	1-2	2-3	3-5	--
On-Land Vacuum	Vert. Wicks	4.5*	3-5	0.5-1	3-6	1
	Horiz. + Vert. wells @	4.5-12	0.5-1	0.5-1	1-2	2
Underwater Vacuum	Horiz. + Vert. wells @	4.5-12	0.5-1	0.5-1	1-2	3

Note: @ = engineered fill with horizontal drains; * = requires 1.5 m of surcharge fill, preload project duration is based on 200-acre site preloaded in multi-stages in different phases.

Performance of Prefabricated Band Drains

Potential for closure of drainage flow paths in the plastic core, due to suction pressure, by the filter fabric and reduction in flow capacity had been a concern in vacuum consolidation with vertical band drains.

From a technical standpoint there is no substantial difference between mechanism of flow induced by conventional surcharging and vacuum preloading, except a suction induced apparent increase of the lateral pressure by a maximum of less than 100 kPa exerted in the band drains for on-land applications. For underwater applications this effect could be larger. Practical experience with band drains in traditional surcharging over a large range of lateral pressures and vertical ground deformations of up to 20% indicates good performance of band drains. The concerns with regard to potential reduction in flow capacity in band drains lead to use of different types of drains. French practice is to use a 2-inch diameter slotted PVC pipe to ensure high flow capacity during vacuum consolidation. Experience at Port of Los Angeles, Tianjin Harbor, and Japan indicates good performance with significant amount of flow through the band drain. Therefore band drains are anticipated to perform well during on-land vacuum consolidation for "typical" sites.

Very limited data is available on the maximum vertical strains that a vertical band drain can withstand in the case of vacuum consolidation of fresh fills or slurries where vertical strains up to or beyond 40-50 percent could be anticipated. Use of non-woven geotextile strip drains in place of plastic core appears to be a viable alternative for surcharge preloading (Fowler 1993). However its application for vacuum has not yet been investigated.

In cases where band drains or prefabricated drainage sheets would be used to consolidate freshly deposited hydraulic fills (or slurries) immediately after placement, clogging of drains may pose problems. Experience with horizontal sand drains in clay-sand layered reclamation projects do indicate significant intrusion of fines into the drains requiring large amount of sands to construct effective drainage layers (Tan et al 1992). This problem may be partly due to penetration of sand particles into underlying very soft fine soils due to lack of sufficient bearing. Little experience is available on the field performance of band drains in the presence of mobile fine-grained soils essentially in a slurry form. In the case of Japanese field trials, a significant reduction in water contents was achieved using band drains in essentially a slurry form of soil. The final water contents were still relatively high. The reason for such high final water contents and the potential effects of clogging are not clear. Potential clogging is a concern that requires careful attention before full scale use of such drains in hydraulic fills or slurries during construction.

CONCLUSIONS

Vacuum-assisted and underpressure consolidation is an effective alternative for conventional surcharging to improve fine-grained soils and expedite self-weight consolidation. Recent technological advances in the application of geosynthetic materials to civil engineering problems have made this technique attractive and cost-effective today. On-land applications are most suitable for soft soil sites with shallow groundwater level. Presence of stratified soils can render vacuum consolidation ineffective unless deeper vertical cut-off-systems are installed. Recent field trials also indicate that on-land vacuum consolidation combined with dewatering can be an effective solution to further accelerate consolidation process.

Experience from on-land field applications of this technology indicate a high potential for use of vacuum technology for improvement of existing hydraulic fills, strengthening weak sediments in the sea floor adjacent to or beneath waterfront retaining facilities, and consolidation of fine-grained hydraulic fills during construction. The technology holds technical merit, and can be cost-effective compared to other conventional techniques. In new land reclamation works the benefits of vacuum consolidation can be realized by inclusion of prefabricated horizontal drains and selective placement of dredge materials. Underwater application or on-land application with dewatering (Fig.2) appears to be most beneficial in such cases. Lack of performance data on prefabricated drains as well as field trials directly applicable for such cases appears to limit its potential uses for many major land reclamation projects at present.

ACKNOWLEDGMENTS

The authors wish to thank the Port of Los Angeles and Messrs. A. Birkenbach, Asst. Chief Harbor Engineer, and G. Alberio, Contract Administrator for their support and contributions.

APPENDIX. REFERENCES

Choa, V., (1989). "Drains and vacuum preloading pilot test," Proc. 12th ICSMFE, Rio de Janeiro, 1347-1350.

Cognon, J.M. (1991). "Vacuum consolidation", Revue Francaise de Geotechnique, No.57-(3), 37-47.

Cognon, J. M., Juran, I., and Thevanayagam, S. (1994) "Vacuum consolidation technology - principles and field experience", Geotech. Spec. Publ. 40(2), ASCE, New York, 1237-1248.

Fowler, J. (1993) personal communication with S. Thevanayagam.

Halton, G.R., Loughney, R.W., and Winter, E., (1965). "Vacuum stabilization of subsoil beneath runway extension at Philadelphia International Airport," Proc. 6th ICSMFE, Montreal, 62-65.

Holtz, R.D., and Wager, O. (1975). "Preloading by vacuum: current prospects", Transp. Res. Rec. 548, TRB, Washington, D.C., 26-29.

Jacob, A., Thevanayagam, S. and Kavazanjian, E. (1994). "Vacuum assisted consolidation of a hydraulic landfill", Geot. Sp. Publ. 40(2), ASCE, New York, 1249-1261.

Johnson, S.J., Cunny, C.W., Perry, E.B., and Devay, L. (1977). "State-of-the-art applicability of conventional densification techniques to increase disposal area storage capacity", Report D-77-4, U.S. Army Engineer WES, Vicksburg.

Kjellman, W., (1952). "Consolidation of clay soil by means of atmospheric pressure" Proc. Conf. on Soil Stabilization, MIT, Massachusetts, 258-263.

Koutsoftas, D.C., and Cheung, R.K.H. (1994) "Consolidation settlements and pore pressure dissipation", Geot. Sp. Publ. 40(2), ASCE, New York, 1100-1110.

Liu, Yi-Xong, (Undated). "History and present situation of soft soil foundation treatment in Xingang, Port of Tianjin," Research Report, First Navigation Engineering Bureau, Ministry of Communications, Tianjin, China.

Port of Los Angeles, (1987). "2020 OFI Study: cargo handling operations, facilities, and infrastructure requirements," Prepared by *Vickery-Zackerman-Miller* for the Port of Los Angeles, Los Angeles, CA.

Shinsha, H., Watari, Y., and Kurumada, Y., (1991). "Improvement of very soft ground by vacuum consolidation using horizontal drains," Proc. Int. Conf. on Geotech. Eng. for Coastal Development (Geo-Coast 91), Port and Harbor Research Institute (Editor), Vol.I, Yokahama, Japan, 387-392.

Stark, T.D., and Fowler, J. (1994) "Management of Dredged Material Placement Operations", Geot. Sp. Publ. 40(2), ASCE, New York, 1353-1365.

Tan, S.A., Liang, K.M., Yong, K.W., and Lee, S.L. (1992) "Drainage Efficiency of Sand Layers in Layered Clay-Sand Reclamation", ASCE, J. Geotech. Eng., Div., 118(2), 209-228.

Tanaka, H., (1994) "Consolidation of Ground Reclaimed by Inhomogeneous Clay", Geot. Sp. Publ. 40(2), ASCE, New York, 1262-1273.

The Earth Technology Corporation (TETC), (1990). "Geotechnical investigation for the Pier 300 42-acre landfill ground modification project, volumes I and II," Report Submitted to the Port of LA, Los Angeles, CA, May.

Thevanayagam, S. (1993). "Prospects of vacuum assisted geosystems for ground improvement at Pier 400", *Internal Report*, prepared for Port of LA, CA.

Woo, S.M., Moh, Z.C., Van Weele, A.F., Chotivittayathanin, R. and Trangkarahart,T.(1989) "Preconsolidation of soft Bangkok clay by vacuum loading combined with non-displacement sand drains", Proc. 12th ICSMFE, Rio de Janeiro, 1431-1434.

Soil Mix Walls in Difficult Ground

David S. Yang[1] M. ASCE and Shigeru Takeshima[2].

ABSTRACT

This paper presents three case histories that illustrate the application of Soil Mix Walls for groundwater control and excavation support in difficult ground where conventional methods were considered difficult to implement technically and economically. The applications include: 1) seepage control through permeable spillway embankment and glacial deposits, 2) groundwater control and excavation support for basement construction in coral ridge and coralline deposits, and 3) groundwater control and excavation support for cut-and-cover tunnel construction in marine clay and glaciomarine deposits.

INTRODUCTION

The SMW (Soil Mix Wall) technology is an in-situ soil mixing technology for soil improvement within a well-defined zone. It consists of mixing in-place soils with cement grout or other reagent slurries using multiple axis augers and mixing paddles to construct overlapped soil-cement columns (Figure 1). Additional connecting and overlapped column panels are then installed to form a continuous subsurface soil-cement wall for use as a cutoff wall (Yang, et.al., 1993) or an excavation support wall (Taki and Yang, 1991). By arranging the column panels in various configurations, this technology becomes a versatile tool for the stabilization of soft or liquefiable soils (Pujol-Rius, et.al., 1988). Using chemical reagents as slurries, this technology has also been used for in-situ solidification and stabilization of contaminated soils (Henderson, et.al., 1993).

---

[1]Senior Engineering Manager, and [2]Vice President of S.M.W. SEIKO, INC., 2215 Dunn Road, Hayward, CA 94545

(a)

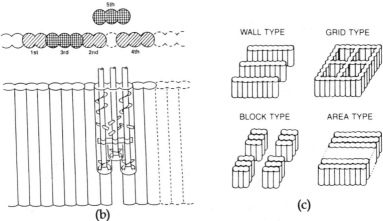

Figure 1: (a) SMW Soil Mixing Equipment, (b) Soil Mix Wall Installation Sequence, (c) SMW Basic Soil Treatment Patterns

Three case studies are presented to illustrate the applications of SMW walls in glacial deposits in Washington, coral deposits in Hawaii, and Boston Blue Clay and glacial marine deposits in Massachusetts. Substantial design and construction details are given in the first case study to provide background on the design, construction, and evaluation processes of the SMW walls. The second and third case studies are presented in a more concise form.

CASE STUDY 1: LAKE CUSHMAN SPILLWAY, WASHINGTON

Background:

In conjunction with the installation of a new radial gated spillway for Lake Cushman near Hoodsport, Washington (Sehgal, et.al., 1992), two sections of embankment were constructed abutting the spillway headworks structure (Figure 2a). The headworks structure was founded on relatively impermeable bedrock. The soil-cement core wall was constructed to bedrock to control water seepage through the embankment fill and the native glacial deposits. The soil-cement core walls were 61 centimeters thick and 61 meters long and 55 meters long within the right and left embankments, respectively. The maximum depth was 43 meters. The core wall material had an average unconfined compressive strength value of 2254 KPa (23 kg/cm^2) and permeabilities on the order of 1×10^{-6} cm/sec (Cotton and Butler, 1990) the core wall profile is shown in Figure 2b.

Subsurface Materials:

The site is underlain by glacial deposits consisting of recessional outwash, lacustrine deposits, and lodgement till. The glacial deposits are underlain by submarine deposited basalt of crescent formation.

The recessional outwash materials consist of dense to very dense fine to coarse sand with trace to little silt, gravelly sand, and sandy gravel with little silt. The lacustrine deposits consist of stiff to very stiff clayey silt, silt, and medium dense to very dense sand with occasional drop stones. The lodgement till consists of very dense gravelly sandy silt, silty sand and gravel with N-values of 50 blow counts for less than 15 cm of sampler penetration. This lodgement till contains cemented zones, cobbles, and boulders, and has been heavily over-consolidated. The basalt bedrock is medium strong to strong with little or no weathered zone. The permeability of the glacial deposits ranged from 10^{-2} to 10^{-5} cm/sec based on the results of seepage analysis, slug tests and packer tests. The permeability of the fresh bedrock ranged between 10^{-6} to 10^{-7} cm/sec based on packer tests.

Two short sections of embankment abutting the spillway headworks were

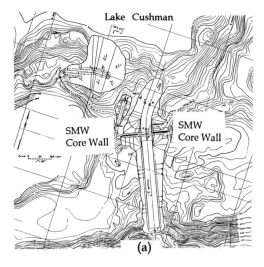

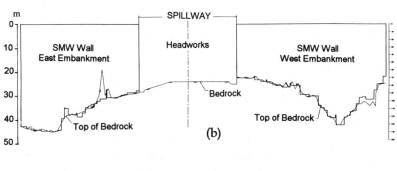

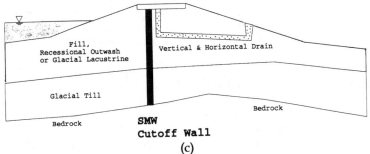

Figure 2: Cushman Spillway Project: (a) Site Plan; (b) Cutoff Wall Profile; (c) Simplified General Cross Section of Embankment

constructed on competent basalt bedrock following the excavation of glacial deposits. The embankment fills consist of compacted silt, sand and gravel.

Design of Soil-Cement Core Wall

Due to the lack of low permeability materials at the site for the construction of a clay core in the embankment and the high permeability of the glacial outwash on the left and right abutments, a soil-cement core wall was designed within the embankment and the glacial outwash to provide for seepage control of water from Lake Cushman. Further investigation revealed that the glacial till contained less fines and was more permeable than expected. Therefore the core wall had to be extended to a maximum depth of 43 meters to reach the bedrock for seepage cutoff as shown in Figure 2c.

To prevent leakage along the interface of the headworks structure and the embankment, a joint structure as shown in Figure 3 was constructed to connect the concrete spillway headworks and the SMW wall. A bentonite and sand mixture was backfilled and compacted within the U-shaped zone during the embankment construction. The core wall was then installed inside the U-shaped zone to form a low permeability joint zone.

Preconstruction Test Section

An 3.4 m long, 4.6 m deep soil-cement core wall test section was constructed in the highly permeable glacial deposits before full-scale wall installation. Three different mix designs selected from laboratory trial mix tests were used. Unconfined compressive strength values of samples retrieved from the soil-cement columns ranged from 706 to 1225 KPa (7.2 to 12.5 kg/cm^2). The test wall was excavated for inspection. Based on the strength test results and observed conditions of the exposed test wall, the owner and the design engineers concluded that the mix designs and the installation method had achieved the design requirements for the planned core wall (Butler and Cotton, 1990).

Construction of Soil-Cement Core Wall

The SMW equipment was placed on top of the spillway embankment and drilled through the embankment, glacial deposits and weathered rock to reach competent basalt. The operation included: 1) working on top of the narrow embankment, 2) installing the wall in permeable embankment and outwash materials, 3) hard drilling in cemented lodgement till containing cobbles and boulders, and 4) reaching competent bedrock which varied in depth as shown on the cutoff wall profile (Figure 2b).

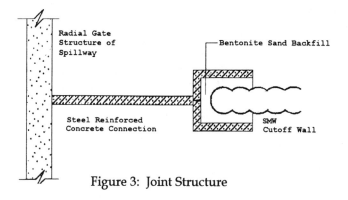

Figure 3: Joint Structure

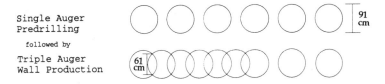

Figure 4: Installation Procedure

The SMW drilling rig is designed to install soil-cement panels outside the tracks of the crawler base machine or between its two tracks. Consequently, it could install the soil-cement core wall on a 6.2 m wide narrow dike (Yang, et.al., 1993). The working pad on top of the spillway embankment was 11 to 13 meters in width which provided sufficient space for the soil-cement wall installation along a core wall alignment located 2 meters from the upstream shoulder of the embankment. The wall was installed panel by panel, therefore, there was no deep open trench inside the embankment to cause embankment stability concerns.

During the wall installation, the bore holes were constantly filled with soil-slurry or soil-grout mixture with a unit weight of 1.6 to 1.7 tons/m^3. This soil-slurry or soil-grout mixture maintained the stability of the bore holes and prevented the sloughing of loose soils from the side surface of the bore holes. These mixtures also formed a filter zone which separated the soil-cement columns and the sandy in-situ soils.

To overcome the hard drilling, predrilling with a single auger was performed and was then followed by the core wall production using the three axis augers as shown in Figure 4. In several locations, large boulders in the lodgement till prevented the predrilling auger from reaching bedrock. By-pass sections were installed around the boulders to maintain the continuity of the core wall.

In the early stages, borings were drilled along the predrilled alignment of the core wall to confirm that the predrilling had reached bedrock. In some instances, bedrock fragments could be retrieved from drill bits after they were withdrawn from a hole during predrilling. Based on this information, field methods were developed to confirm drilling to the top of the bedrock with irregular depths (Figure 2b).

Soil-Cement Core Wall Constructed

Mix designs with cement dosages of 350 to 550 kg per cubic meter of in-situ soil were used for core wall installation. The west wall consisted of 1733 m^2 of soil-cement core wall with a 28-day unconfined compressive strength ranging from 594 to 4,396 KPa (6 to 45 kg/cm^2) and permeabilities ranging from 2×10^{-5} to 6×10^{-7} cm/sec. The east wall consisted of 2,219 m^2 of soil-cement core wall with strength ranging from 1800 to 4790 KPa (18 to 49 kg/cm^2) and permeabilities ranging from 1×10^{-6} to 7×10^{-7} cm/sec. The core walls were constructed in compliance with the design specifications.

CASE STUDY 2: MARIN TOWER PROJECT, HAWAII

Background:

Marin Tower, a mixed-use project owned by the City and County of Honolulu, consists of a 28 story tower with a two level subsurface parking garage in highly permeable coral ridge and coralline deposits. Located approximately 30 meters from the harbor, a large quantity of water was expected to pass through the void in these soils, making dewatering difficult during basement construction. The in-situ soil mixing method was selected to create cutoff walls for groundwater control. Three sides of the soil-cement wall were reinforced for use as excavation support in conjunction with tiebacks.

Subsurface Materials:

The site is underlain by coral reef, reef detritus materials, and alluvium deposits. The three meter thick upper coral reef consists of hard coralline limestone with cavities. The reef detritus materials consist of medium dense sandy coral gravel interbedded with layers of loose to medium dense coral sand and occasionally with thin layers of alluvium. These detritus materials extend to depths varying from 4 to 35 meters. The upper alluvium deposits consist of clayey silt and sand layers generally encountered between 15 to 24 meters and locally interbedded with reef detritus. The lower alluvium deposits consist of layers of sand, basalt gravel, basalt boulders, and cobbles, and extend to the maximum exploration depth of 47 meters except for one boring location where a basalt rock formation was encountered at a depth of 40 meters.

Groundwater Level:

The observed groundwater level varied from three to six meters below the ground surface. Due the proximity of the site to the ocean, the groundwater level fluctuates with tidal changes.

Selection and Design of SMW Wall:

Sheet pile walls and soldier piles and lagging have been used in the Honolulu area for excavation support. At sites with high groundwater tables, dewatering or sheet piles have been used. Recent developments of water front structures with basements encountered difficulties due to soft subsurface soils and the large volume of groundwater in the highly permeable or porous coral formation or coralline deposits. Unlimited pumping during construction caused distress to adjacent structures. To control the horizontal flow of groundwater into the excavation, sheet pile walls were installed in predrilled holes filled with lean concrete. This is a difficult construction operation, suffering from both inefficiency and high cost. At the Marin Tower site, the soil-cement wall together with the sheet pile wall, and soldier piles and lagging in conjunction with dewatering were evaluated as alternatives. The soil-cement wall was selected based on: 1) effective control of lateral groundwater flow, 2) no need for predrilling or driving to penetrate the upper hard coral reef, 3) no subsidence or distress to adjacent structures, 4) short construction time, and 5) relatively low cost.

Design of Soil-Cement Wall:

The ideal application of soil-cement walls is to key the wall into the low permeability stratum below the bottom of the excavation for control of groundwater in both the horizontal and vertical directions. The soil strata vary significantly across the site and a reliable and continuous low permeability layer could not be located near the bottom of the excavation. Therefore, a partial cutoff scheme was developed by extending the soil-cement wall to an average depth of 14 meters to control the lateral groundwater flow and to reduce the amount of vertical flow through the increased length of the seepage path. H-piles were included in every alternate soil-cement column to resist lateral pressure. For toe stability, a 3.4

meters embedment was required and this portion of the soil-cement wall was reinforced with H-piles. An additional three meters of soil-cement wall below the toe of the H-piles was not reinforced and served only as a seepage cutoff wall.

Construction of Soil-Cement Wall:

A 55 cm diameter three axis auger was used to drill through the coral reef. No predrilling was required. The drilling speed was adjusted to break and grind the coral limestone into gravel size or smaller for thorough mixing with cement grout. The soil-cement mixture thus produced consisted of particles with sizes ranging from gravel to silt and filled the entire depth of the bore holes and the overflow control trench. The unit weight of the mixture was approximately 1.6 kg per liter (100 pounds per cubic foot) and therefore the pressure inside the bore holes or soil-cement panels was higher than the hydraulic pressure and prevented the water from entering the soil-cement panels. When cavities were encountered in the coral reef, the soil-cement mixture from above and from neighboring panels flowed in to fill and stabilize the cavities and prevented further loss of soil-cement mixture.

Soil-Cement Wall Constructed

Mix designs with cement dosages of 300 to 500 kg per cubic meter of in-situ soil were used for the soil-cement wall installation. A total of 4,015 square meters of soil-cement wall was constructed. Eighty percent of the wall served the dual function of excavation support and groundwater control. The soil-cement wall parallel to the underpinned existing structure was not reinforced and only served as a cutoff wall. The 28-day unconfined compressive strength ranged from 833 to 1431 KPa (8.5 to 14.6 kg/cm^2). A view of the soil-cement wall is shown on Figure 5. The circular surface of soil-cement columns was chipped off using a backhoe to create a flat surface for the placement of volclay sheets and construction of the permanent basement wall. The horizontal flow of water through the soil-cement wall was negligible. Due to the absence of a low permeability layer near the bottom of the excavation, a total of eleven dewatering wells were installed and five were constantly used to control the bottom flow of groundwater. The pumping rates ranged from 2,000 to 6,000 gallons per minute. Horizontal flow of groundwater was effectively controlled by the soil mix wall, enhancing the efficient construction of the basement.

Figure 5: SMW Wall, Marin Tower Project

Figure 6: SMW Wall, Central Artery/Tunnel Project

CASE STUDY 3: BOSTON CENTRAL ARTERY/TUNNEL C07 PROJECT, MASSACHUSETTS

Background

As part of the Third Harbor Tunnel project, a cut-and-cover tunnel was designed to connect the Logan International Airport and the immersed twin steel tube tunnel crossing the Boston Harbor. The excavation cut is 1,128 meters long with the depth ranging from 12.5 to 26 meters. The majority of the cut sections were supported by a soil mix wall and a short section where the cut-and-cover tunnel is connected to the twin tube tunnel was supported by a concrete slurry wall. Two soil-mix walls, a 400 meter long east wall and a 500 meter long west wall were installed using 91 cm diameter three-axis augers. The soil mix wall was designed and constructed by a joint venture of Nicholson Construction Company and S.M.W. SEIKO, INC.

Subsurface Materials:

The subsurface conditions in the cut-and-cover tunnel were influenced by a complex historical site filling and various past site uses. The subsurface materials in descending order consisted of fill, organic deposits, marine deposits, glaciomarine deposits, glaciolacustrine deposits, glaciofluvial deposits, glacial till deposits, and bedrock (Davidson, et.al., 1991) Some strata may not exist in certain sections of the tunnel alignment.

The fill material consists of granular fill, miscellaneous fill and cohesive fill. The granular fill is typically fine sand, with little fine gravel, trace silt, cinders, brick, wood, and concrete debris. The miscellaneous fill is typically sand and silt with sand, gravel, cobbles, organic material, clay, cinders, brick, wood, and concrete. The cohesive fill consists of clay, organic silt, and silt intermixed with sand, gravel, wood, and shells.

The organic deposits were deposited in tidal marsh and estuarine environments and consist of organic silt with sand, silt and clay. The thickness varied from one to three meters when encountered.

The marine deposits along the tunnel alignment consist mainly of gray clay, known locally as "Boston Blue Clay," with lenses of sand and silt.

The glaciomarine deposits can be divided into two subunits: upper glaciomarine and lower glaciomarine. The upper unit consists of silt with little sand, clay, gravel and cobbles. The lower unit consists of silt with little gravel, sand, clay, cobbles, and boulders. The primary distinctions between the two units are the cohesion and N-values. The upper unit exhibits greater cohesion. The lower unit has higher SPT N-values and closely resembles the very dense glacial till.

Design of Soil Mix Wall

The soil mix wall at the site was reinforced with W21x50 H-piles used at 1.2 meter (four feet) intervals in conjunction with four to five tiers of anchors for excavation support. The wall also functions as a cutoff wall for groundwater control. The soil-cement transfers lateral earth pressure and hydraulic pressure to the neighboring H-piles. The H-piles were designed to carry the lateral pressure as conventional soldier piles do. An unconfined compressive strength of 620 KPa (6.3 kg/cm^2) for the soil-cement was designed to resist the stresses during lateral pressure transfer.

A laboratory trial mix study was performed on three major types of soil samples. Based on the test results, several mix designs were selected for the soil mix wall construction in different subsurface soils.

Construction of Soil-Mix Wall

Subsurface conditions along the tunnel alignment are complex and vary significantly (Davidson, et.al., 1991). The installation procedures of the soil mix wall were varied to cope with the different subsurface conditions. The subsurface conditions which affect the wall installation are grouped into three distinct zones for discussion of the wall installation.

Zone A - Thick Glaciomarine Deposit: The wall in this zone was extended to a maximum depth of 28 meters. The subsurface soils encountered during the soil mixing were 4 to 6 meters of granular fill and approximately 18 to 22 meters of glaciomarine with lenses of organic deposits and Boston Blue Clay between the two major strata. The SPT N-values ranged from 30 to 60 blows per foot in the upper glaciomarine and, in general, in excess of 100 blows per foot in the lower glaciomarine. The selection of the wall installation procedure was based on the conditions of drilling and soil mixing in the glaciomarine. Due to the cementation and the boulders in this stratum, predrilling using a single auger was planned for penetration into the very dense glaciomarine along the whole alignment in this zone. However, predrilling was found to be unnecessary in most of this section and the three-axis auger was use for wall production directly with satisfactory penetrability and a satisfactory level of soil mixing.

Zone B - Thick Fill Deposits: The wall in this zone was extended to a maximum depth of 25 meters. This section was an early 1900's shipping channel which was later filled. The depth of fill encountered was typically

15 to 18 meters. The fill is a combination of granular fill, miscellaneous fill and cohesive fill. Numerous obstructions including concrete debris, wood, rock fragments, and boulders were encountered making the installation of the continuous wall by the three-axis auger very difficult and inefficient. Therefore, predrilling using a single auger at four feet spacing was used to break the obstructions and to create a passage for the three-axis auger wall production. In areas where there were large quantities of concrete debris and boulders, the single auger predrilling was also difficult. Therefore, obstructions were removed by excavating with a backhoe. The excavation was then backfilled with granular material for wall production.

Zone C - Thick Marine Deposits: A section of the tunnel alignment, approximately 260 meters long is underlain by a deep trough filled with Boston Blue Clay to a maximum depth of 30 meters. This marine clay is very stiff to hard at the upper three meters and the lower portion ranges between medium stiff to stiff. The atterberg limits range from 25 to 55 with plastic limits between 20 to 25. The natural water content generally ranges from 30 to 40 percent. The marine clay is overlaid by six meters of fills and one to two meters of organic deposits. The procedures for soil mixing and wall installation were determined by the properties of the dominant material, Boston Blue Clay. A large quantity of grout was used to break and blend the highly plastic clay for soil mix wall production.

Soil Mix Wall Constructed

A total 37,180 square meters of soil mix wall (Figure 6) was constructed for excavation support and groundwater control. For transfer of lateral pressure to the neighboring H-piles, an unconfined compressive strength of 620 KPa (6.3 kg/cm^2) was required. Due to construction difficulties at the site, various installation procedures were used in three distinct zones to obtain required levels of soil mixing while maintaining the construction schedule. The strength tests results along the alignment of the tunnel are summarized in Figure 7 together with cement dosage.

SUMMARY

Three case studies were presented to illustrate the construction of soil mix walls in difficult ground including 1) cemented glacial till and glacial outwash in Washington; 2) coral reef and coral deposits in Hawaii; and 3) glaciomarine, Boston Blue Clay, and fill with concrete debris and large boulders in a previous ship channel in Massachusetts.

The applications illustrated included: 1) seepage control inside a spillway embankment and high permeability glacial outwash; 2) excavation support

SOIL MIX WALLS

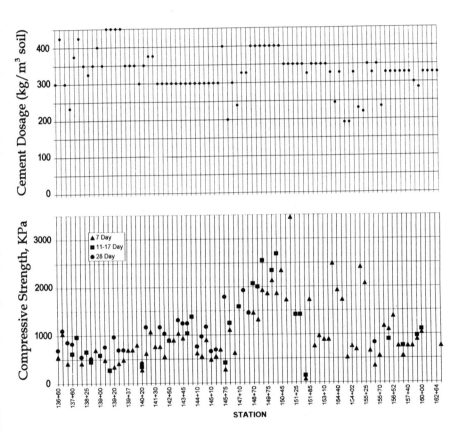

Figure 7: Strength and Cement Dosage

and groundwater control in coral deposits for basement construction; and 3) excavation support and groundwater control for cut-and-cover tunnel construction in heterogeneous subsurface soils. The in-situ soil mixing technology provides a versatile approach to modify difficult subsurface soils to obtain required engineering properties for various construction purposes.

REFERENCES

Cotton, D.M, and Butler, S. (1990), "Summary Report for Soil-cement Core Wall Installation," Golder Associates, Inc., Redmond, WA.

Davidson, W.A.; Gifford, D..G.; and Kraemer, S.R. (1991). "Final Geotechnical Engineering Report, Design Sections D007A and D007C, Central Artery (I-93)/Tunnel (I-90) Project," Haley & Aldrich, Inc. Cambridge, MA.

Henderson, Lisa; Delfino, Thomas; Rafferty, Michael; Kimura, Tetsuo; Takeshima, Shigeru; Lovell, Douglas; and Yang, David; (1993), "In-situ Fixation of Arsenic and Heavy Metals in Soils," Summer National Meeting, American Society of Chemical Engineers, Seattle, WA.

Pujol-Rius, A.; Griffin, P.; Neal, J.; and Taki, O. (1989), "Foundation Stabilization of Jackson Lake Dam," 12th International Conference on Soil Mechanics and Foundation Engineering, Brazil.

Sehgal, C.K.; Fischer, S.H.; and Sabri, R.G; (1992) "Cushman Spillway Modification," USCOLD Conference, Fort Worth, TX.

Taki, O. and Yang, D.S., (1991), "Soil-Cement Mixed Wall Technique," ASCE Special Conference, Denver, CO.

Wong, B.Y.K. (1990), "Preliminary Geotechnical Engineering Exploration, Smith Maunakea Housing," C. W. Associates, Inc. and DBA Geolabs-Hawaii, Honolulu, Oahu, HI.

Yang, D. S.; Luscher, U.; Kimoto, I.; and Takeshima, S.; (1993), "SMW Wall for Seepage Control in Levee Reconstruction," Third International Conference on Case Histories in Geotechnical Engineering, St. Louis, MO.

Control of Settlement and Uplift of Structures Using Short Aggregate Piers

By Evert C. Lawton,[1] Member ASCE, Nathaniel S. Fox,[2] Associate Member ASCE, and Richard L. Handy,[3] Fellow ASCE

ABSTRACT: Two case histories are described in which short aggregate piers were used to control settlements and uplift movements. In one project, aggregate piers were effective in substantially reducing settlements of both individual footings and mat foundations. A method for estimating settlements of the aggregate pier-reinforced soil is described, with good correlation shown between predicted and actual settlements. Aggregate piers were used in a second project to provide uplift capacity for an airplane hangar susceptible to high uplift wind forces. The aggregate piers have performed well in winds up to 113 km/hr (70 mi./hr). A theoretical method for estimating the uplift capacity of aggregate piers is proposed.

INTRODUCTION

Frustrated by limitations posed by the overexcavation and replacement method of stabilizing poor or inadequate soils to support footings and control settlements, an effort was begun in 1984 to develop a more practicable, higher capacity method of providing soil reinforcement to support shallow foundations. The method developed was a short aggregate pier system involving the formation of a cavity by drilling or backhoe excavation, and building a highly densified, well-graded aggregate pier in lifts by impact ramming, while simultaneously causing buildup of lateral and vertical soil stresses. Projects utilizing this system since 1988 have included a variety of structures on widely differing soil conditions, In the past three years, with improved apparatuses and installation techniques, several thousand aggregate piers have been installed to support a wide variety of structures by providing settlement control, and in several projects, uplift and horizontal movement control. The system has been expanded to include stiffening and strengthening of good soils. The unique differences inherent within the aggregate pier system compared to other foundation types or ground improvement methods have resulted in the award of a U. S. patent (Fox and Lawton 1993), with international patents pending.

BACKGROUND

The aggregate pier method has been developed as an economical alternative to overexcavation/replacement and to deep foundations in many instances. Its primary use to date has been to control settlements beneath building footings and mats, while providing higher capacity, higher bearing pressure foundation elements. Projects

---

[1] Assoc. Prof., Dept. of Civil Eng., Univ. of Utah, 3220 MEB, Salt Lake City, UT 84112.
[2] President, Geopier Foundation Co., Inc. 769 Lake Drive, Lithonia, GA 30058.
[3] Distinguished Prof. Emeritus, Iowa State Univ., Dept. of Civil Engrg. Ames, IA 50011.

completed range from single story structures to sixteen story towers, and from storage silos to airplane hangars. Soils improved with high capacity aggregate piers have included soft and loose sandy silts and silty sands, soft and firm silty clays and clayey silts, organic fills, debris fills, uncompacted or erratically compacted fills, very stiff silts, and medium dense sands. Groundwater conditions have varied from none to groundwater existing within the aggregate pier elements. Aggregate piers have been used on several projects to control uplift. Details of three projects in which aggregate pier foundations were used to control settlements have been described previously (Lawton and Fox 1994).

The major steps involved in creating an aggregate pier within an in-situ (matrix) soil are illustrated in Fig. 1 and summarized as follows: (a) A cylindrical or rectangularly prismatic cavity is formed in the soil using either an auger or a backhoe; (b) the soils at the bottom of the cavity are densified and prestressed by repeated impact from a specially designed tamper with a beveled head; (c) well-graded aggregate (normally highway base course stone) is placed loosely at the bottom of the cavity in a thin lift; (d) the aggregate is highly densified (typically to more than 100% modified Proctor maximum dry density) by repeated ramming from the tamper, which also prestresses the matrix soil laterally; and (e) compacted lifts are added until the desired height is achieved. Columnar aggregate piers have varied in diameter from 0.61 to 0.91 m (24 to 36 in.), while the rectangularly prismatic piers have varied in width from 0.46 to 9.76 m (18 to 30 in.). The height of an aggregate pier is generally between two to three times its diameter or width. The apparatuses used to densify the aggregate have included a small skid loader with a modified hydraulic impact source, a backhoe, and a hydraulic excavator. The diameters of the tamper heads have varied from 0.38 to 0.81 m (15 to 32 in.).

CASE HISTORIES

Expansion of Regional Hospital, Atlanta, Georgia

In a $312,000,000 expansion to a regional hospital in Atlanta, Georgia, two major structural towers were added to provide additional office space and hospital rooms. One tower was designed to add twelve stories on top of an existing four story building component, providing a total height of sixteen stories, while the other tower was to bear at the basement level and extend a full sixteen stories. Although each tower was designed with drilled pier foundations, there were distinct foundation installation problems associated with drilled piers for each case.

Subsurface and Construction Conditions, South Tower. The site is within the Piedmont geological province of Georgia. Subsoils in both tower areas, which are within 91 m (300 ft) of each other, were virgin residual soils consisting primarily of firm to very stiff fine sandy micaceous silt and loose to firm silty micaceous sands overlying dense silty micaceous sands and partially weathered rock. The total thickness of the soil strata overlying partially weathered rock and rock varied from about 3.0 to 9.1 m (10 to 30 ft). Standard penetration blowcounts (N) in the soils varied from as low as 4 to above 20, and typically ranged from 9 to 20. The underlying rock was classified as soft to hard gray and white biotitic gneiss. Groundwater at the time of drilling was found at elevations 0.9 m (3 ft) above finished floor elevation in one boring, and 1.2 to 4.6 m (4 to 15 ft) below finished floor elevations in other borings.

The foundations for the South Tower (twelve story) had to be installed within an existing basement. The ceiling height of 4.9 m (16 ft) presented one limitation. Access to the site was limited to a wall opening about 2.4 m (8 ft) wide and 3.0 m (10 ft) high. Furthermore, access to the opening was by a steep ramp with a 1.3H:1.0V slope. Alternatively, equipment could be lifted down to the construction

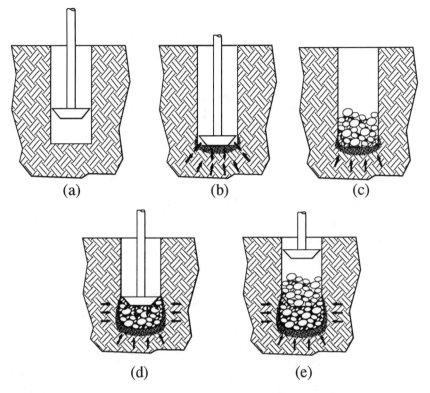

FIG 1. Steps in Construction of an Aggregate Pier

level by crane. Because of the equipment height and size limitations, hand-dug drilled piers were initially planned.

The initial plan to install 0.91 m (36 in.) diameter columnar piers was modified when it was learned that the lifting crane could not handle the drill rig. Rectangularly prismatic (linear) aggregate piers with widths of 0.61 m (24 in.) were therefore substituted for the columnar piers. The areal coverage of the aggregate piers was about one-third of the total footing area. The aggregate piers were installed by excavating a trench about 1.83 m (6 ft) below the bearing elevation and compressing the soils at the bottom of the trench to a depth of about 0.15 m (6 in.), resulting in a total height of the piers of about 1.98 m (6.5 ft).

Subsurface and Construction Conditions, North Tower. The foundations for the North Tower (sixteen stories) had to be installed within a 6.1 m (20 ft) deep basement excavation braced with a tieback retention system. Access to the basement was provided by an access ramp with about the same slope as for the South Tower. Drilling equipment for the drilled piers was to be lifted by crane onto and off of the site.

The geology and subsoil conditions were essentially the same as for the South Tower with the following exceptions: (1) Groundwater existed 0.37 m (18 in.) below the ground surface of the basement excavation; (2) the depth to rock was greater; and (3) results of dynamic penetration tests on the subsoils showed that the lower consistency sandy silts and silty sands extended only approximately 1.5 m (5 ft) in depth and that the underlying subsoils were stiff to very stiff and firm to medium dense below this depth.

The 24 hour time period immediately prior to installation of the aggregate piers produced a record rainfall of 102 mm (4.0 in.). The site, being a deep excavation adjacent to the existing hospital, offered no drainage except seepage into the subsoils. As a result, the upper subsoils were saturated to varying depths, resulting in "pumping" of both wheeled and tracked construction equipment. To stabilize these soils to support wheeled skid loader tamping and front loading equipment, and to provide protection against trench cave-ins resulting from wheel loads near the edge of the pier excavations, the authors elected to install geogrid on the surface with 102 mm (4 in.) of #57 stone on top. This performed well, stopping the subgrade pumping and preventing cave-ins during installation.

Because of strict schedule constraints and the threat of additional hard rains, the installation of the aggregate piers was continued non-stop using two 12-hour shifts. The total project was completed within a 48 hour time frame. Several hours after completion, the predicted heavy rainfall occurred. The highly densified aggregate piers were not degraded by the action of this rain.

Settlement Analyses. Idealized geologic profiles for the South Tower footings and the North Tower mat are shown in Fig. 2, along with pertinent engineering properties of the strata. In the South Tower, structural loads for the columns supported on aggregate pier-reinforced soils varied from 1.69 to 4.40 MN (380 to 990 kips). The aggregate pier system was designed to provide a maximum design bearing pressure of 5.0 ksf, with resultant footing sizes ranging from 2.74 m (9 ft) square to 3.66 by 5.03 m (12 by 16.5 ft). The foundation system for the North Tower was originally designed as isolated columns and grade beams supported by drilled piers. This system was replaced by an aggregate-pier supported mat with an average bearing pressure of about 144 kPa (3.0 ksf), and a maximum bearing pressure within the heavier loaded portion of about 215 kPa (4.5 ksf).

To estimate the settlement of a shallow foundation bearing on an aggregate pier-reinforced soil, the subgrade is divided into an upper zone (UZ) and a lower zone (LZ). The upper zone is assumed to consist of the composite soil comprised of the aggregate piers and matrix soil, plus the zone of appreciable densification and prestressing immediately underlying the pier, which is estimated to be equal to the width of one pier. For this case, the piers were 1.98 m (6.5 ft) high and 0.61 m (24 in.) wide, with the height of the upper zone equal to 2.59 m (8.5 ft). The lower zone consists of all strata beneath the upper zone. Settlements are calculated individually for the UZ and LZ, with the two values combined to yield an estimate of the total settlement. Using the analyses to be discussed subsequently, predicted settlements were calculated for both the smallest and largest footings in the South Tower. The mat for the North Tower was 15.2 by 30.5 m (50 ft by 100 ft), with about 75% of the mat loaded to 144 kPa (3.0 ksf) and about 25% loaded to 215 kPa (4.5 ksf). Settlement of the mat was calculated separately for the two distinct zones with different applied loads. Since rock was so close to the surface (ratio of soil thickness to width of the loaded area was small), varying the assumed size of the different zones of the mat had little influence on the predicted settlements. Therefore, settlements of the lighter loaded zone were calculated based on the full

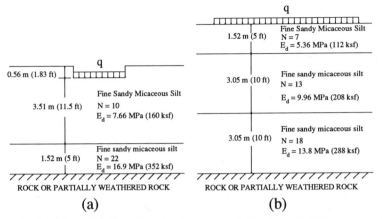

FIG. 2. Idealized Geologic Profiles for Hospital Project: (a) South Tower; (b) North Tower

dimensions of the mat, and settlements of the heavier loaded zone were calculated based on dimensions of 7.6 by 15.2 m (25 by 50 ft).

Over the past six years, the authors have conducted numerous settlement analyses for the UZ using the finite grid method (Bowles 1988), which have shown that little error is introduced in the settlement calculations by assuming that the footing is perfectly rigid. Using this assumption and a subgrade modulus approach, the following equations apply:

q_p = bearing stress applied to aggregate piers = $q \cdot R_s / (R_a \cdot R_s - R_a + 1)$ .... (1)
q_m = bearing stress applied to matrix soil = q_p / R_s .... (2)
S_{uz} = settlement of the UZ = $q_p / k_p = q_m / k_m$ .... (3)

where q = average design bearing pressure = Q / A; R_s = subgrade modulus ratio = k_p / k_m; R_a = area ratio = A_p / A; Q = vertical design load at the bearing level; A = total area of footing; A_p = total area of aggregate piers supporting footing; k_m = subgrade modulus for matrix soil; and k_p = subgrade modulus for aggregate piers.

Values of subgrade moduli for the aggregate piers are determined either by static load tests on individual piers or by estimation from previously performed static load tests within similar soil conditions and similar aggregate pier materials and installation methods. This is considered conservative since the static load tests do not consider the beneficial effect of confining pressures produced from the loaded footing acting on the matrix soil. Subgrade moduli for the matrix soils are either determined from static load tests or estimated from boring data and allowable bearing pressures provided by geotechnical consultants. Because the aggregate piers are typically 10 to 20 times as stiff as the matrix soil, the value of subgrade modulus for the matrix soil generally has only a minor influence on the estimated settlement within the UZ.

For this project, no load tests were conducted on either the aggregate piers or the matrix soil, so values of subgrade moduli for both materials were estimated. From the results of over thirty static load tests on aggregate piers installed within matrix soils from the Piedmont geological region, values of k_p for aggregate piers installed

Table 1. Predicted Upper Zone Settlements

Location and Foundation Type	Plan Dimensions of Foundation	q kPa (ksf)	R_a	q_p kPa (ksf)	q_m kPa (ksf)	S_{uz} mm (in.)
South Tower square footing	2.74 m (9 ft)	225 (4.7)	0.35	563 (11.8)	42 (0.88)	7.4 (0.29)
South Tower rectangular footing	3.66 x 5.03 m (12 x 16.5 ft)	239 (5.0)	0.33	630 (13.1)	47 (0.99)	8.3 (0.33)
North Tower mat	15.2 x 30.5 m (50 x 100 ft)	144 (3.0)	0.29	418 (8.7)	31 (0.66)	5.5 0.22
North Tower mat	7.6 x 15.2 m (25 x 50 ft)	215 (4.5)	0.29	628 (13.1)	47 (0.98)	8.3 (0.33)

Note: For all cases, k_p = 76 MN/m³ (280 pci), k_m = 5.8 MN/m³ (21 pci), and R_s = 13.3.

in similar types of subsoils have varied from 49 to 190 MN/m³ (180 to 700 pci). Based on load tests from sites with similar soils and blowcounts, the k_p for this project was estimated at 76 MN/m³ (280 pci). Using an allowable bearing pressure of 144 kPa (3.0 ksf) based on a tolerable settlement of 25 mm (1 in.), k_m was estimated as 5.8 MN/m³ (21 pci). Using R_s = 13.3 and these values for k_p and k_m, as well as the values for q and R_a shown in Table 1, values for q_p, q_m, and S_{uz} for both footings were calculated from Eqs. 1, 2, and 3 and are summarized in Table 1.

An estimate of the applied stresses transmitted to the interface between the UZ and the LZ is needed so that predicted settlements in the LZ can be calculated. Burmister's (1958) work on two-layered elastic strata of infinite horizontal extent clearly showed that the presence of a stiffer upper layer substantially reduces the applied stresses transmitted to the lower, more compressible layer compared to the case of a homogeneous soil. For example, for a uniform circular load, $E_1 / E_2 = 10$, and a thickness of the upper layer equal to the radius of the loaded area, the vertical normal stress beneath the centerline of the loaded area at the interface between the two materials is about 30% of the applied stress, compared to about 65% for a homogeneous soil (Boussinesq-type analysis) at the same depth.

The procedure used by the authors to estimate vertical stress increase at the UZ-LZ interface is a modification of the 2:1 method, and involves the use of engineering judgment. Use of this type of method is readily applicable to settlement calculations of the LZ because it provides an estimate of uniform vertical stress increase at the UZ-LZ interface. For estimates of the lower zone settlements (S_{lz}) for this project, a stress dissipation slope through the UZ of 1.67:1 is used.

To estimate settlements of the lower zone, and to estimate settlements for comparable footings without aggregate piers (S_{un}), both Schmertmann's (1970; 1978) strain distribution method and Bowles' (1988) modified elastic theory method are used. Based on comparisons of SPT blowcounts in Piedmont residual soils versus Young's modulus values backcalculated from actual settlements and estimated from plate load and other in-situ tests, the authors have found that most values for drained Young's modulus fall within the following range: E_d (kPa) = (383 to 1149)·N [E_d (ksf) = (8 to 24)·N], where N is the field blowcount (not corrected for overburden pressure). Although there is considerable scatter in the data, the best straight-line fit to the data is E_d (kPa) = 766·N [E_d (ksf) = 16·N], and this relationship has proved satisfactory for most settlement estimates in Piedmont residual soils and is used herein. The results of these calculations are summarized in

Table 2. Predicted Settlements for Lower Zones and Unreinforced Matrix Soils

	Predicted Settlement, mm (in.)			
	Lower Zone, S_{lz}		Unreinforced, S_{un}	
Location and Foundation	Schmert.	Bowles	Schmert.	Bowles
South Tower square ftg.	0.8 (0.03)	4.6 (0.18)	43.2 (1.70)	40.4 (1.59)
South Tower rect. ftg.	1.5 (0.06)	7.9 (0.31)	51.8 (2.04)	58.2 (2.29)
North Tower large mat	7.4 (0.29)	32.5 (1.28)	28.7 (1.13)	70.6 (2.78)
North Tower small mat	15.7 (0.62)	39.6 (1.56)	61.5 (2.42)	98.3 (3.87)

Table 3. Predicted and Actual Settlements

	Settlement, mm (in.)		
	Predicted		
Location and Foundation	Unrein. Matrix Soil	with Aggr. Piers	Actual
South Tower sq. ftg.	40 - 43 (1.6 - 1.7)	8 - 12 (0.3 - 0.5)	< 6 (< 0.25)
South Tower rect. ftg.	52 - 58 (2.0 - 2.3)	10 - 16 (0.4 - 0.6)	< 10 (< 0.4)
North Tower large mat	29 - 71 (1.1 - 2.8)	13 - 38 (0.5 - 1.5)	< 10 (< 0.4)
North Tower small mat	62 - 98 (2.4 - 3.9)	24 - 48 (0.9 - 1.9)	< 20 (< 0.75)

Table 2. Both methods gave comparable values for unreinforced settlements of the South Tower footings. In the other six cases, Bowles' method yielded predicted settlement values significantly higher than Schmertmann's method; values calculated for the mats using Bowles' method seem especially high.

The total predicted settlements, with and without aggregate piers, are shown in Table 3 along with the actual settlements. In all four cases, the minimum value of predicted settlement with aggregate piers (using Schmertmann's value for the LZ) is slightly larger than or equal to the maximum value of actual settlement, suggesting that the settlement method used by the authors gives reasonable estimates. Values of predicted settlement with and without aggregate piers indicate that the aggregate piers were effective in reducing both total and differential settlements.

Mississippi Air National Guard Hangar, Meridian, Mississippi

A state-of-the-art hangar with massive doors that fold up, much like venetian blinds, was built at an Air National Guard field in Meridian, Mississippi. Owing to the open door space, design uplift forces from wind loads were as great as 1,156 kN (260 kips) per column. Helical screw anchors were considered and initially bid, but an alternative anchoring system was sought because the helical anchors presented two problems: (1) Difficulty in locating the anchor shafts within specified tolerances; and (2) a significantly higher cost than was previously budgeted.

Since December 1991, the authors have successfully used aggregate piers as holddown anchors during compressive static load tests on aggregate piers, with two significant characteristics observed during tests conducted in silty sands and sandy silts: (1) The uplift capacity of an aggregate pier was significant; and (2) in 31 of 32 piers where uplift deflections were measured, the rebound upon removal of the load was 100%. These results suggest that the uplift loads were transferred primarily as shear stresses along the aggregate pier-matrix soil interface, and that the stresses were within the elastic range for the interfacial materials. The maximum uplift forces per aggregate pier in these tests were typically between 200 to 214 kN (45 to 48 kips), with measured uplift deflections mostly less than 25 mm (1.0 in.) and always less than 51 mm (2.0 in.), and in no cases did failure occur. The results from uplift tests in sandy clays have shown less than 100% rebound, indicating some plastic soil behavior.

Geological Background. The Key Field Hangar site is located within the Coastal Plain geologic region, and identified within the Wilcox formation. The sedimentary soils in this region typically consist of complexly interbedded clays and sands, and are usually underlain by sandstone or limestone. Subsoils within the hangar site included an upper zone of well-compacted, well-graded sand fill extending from the graded surface to depths of about 0.9 to 1.8 m (3 to 6 ft). Available test data indicated that the density of the sand fill was approximately 98% of standard Proctor maximum, with N ≈ 15. Underlying the recent sand fill was a zone of primarily loose clayey sand varying from 1.2 to 2.7 m (4 to 9 ft) thick. Unconfined compressive strengths in the clayey sand varied from 48 to 192 kPa (1.0 to 4.0 tsf), and more typically from 72 to 96 kPa (1.5 to 2.0 ksf). Beneath this was a stratum consisting of stiff fine sandy silt and medium dense silty fine sand extending to a depth of 7.6 m (25 ft). Groundwater was at depths of about 1.5 to 2.1 m (5 to 7 ft), and near or within the bottoms of the installed piers.

The subsoil profile at the location of the uplift load test consisted of 1.4 m (4.7 ft) of medium dense, well-graded sand fill overlying a zone of loose, very clayey sand. The groundwater was 0.3 m (1 ft) above the bottom of the pier excavation, at a depth of about 1.8 m (6 ft). Consolidated-drained strength parameters determined from borehole shear tests performed within the aggregate pier test zone prior to excavating the pier cavity were $\phi = 45°$ and $c \approx 0$ for the compacted sand fill, and $\phi = 20°$ and $c \approx 0$ for the clayey sand.

Uplift Load Test. Rectangularly prismatic aggregate piers were used instead of columnar piers because of subsoil conditions and anticipation of some limited cave-in situations. It was also felt that if the uplift capacity were less than anticipated, greater aggregate pier coverage and depth would be more readily available with backhoe excavation than drilled hole excavation.

The test aggregate pier was 1.8 m (6 ft) high, 0.61 m (24 in.) wide, and 1.5 m (5 ft) long, with the top of the pier at a depth of 0.3 m (1 ft). The uplift loads were transferred to the bottom of the pier by steel tension rods located along the perimeter of the pier, which were attached to a continuous steel plate at the bottom of the pier.

The load test was performed essentially in accordance with ASTM D1194. A total of eight loading increments were applied, with an average increment of 67 kPa (15 kips) each. The time between loading increments was 15 minutes; for each increment the deformation rate after 15 minutes was less than 0.25 mm (0.01 in.) per minute. The maximum load of 267 kN (60 kips) was held for five hours. As can be seen in Fig. 3, the deflection at the maximum load was 23 mm (0.91 in.), and the load-deflection curve is fairly linear. In addition, 100% rebound was measured upon release of the load, indicating that the soil behaved elastically within the range of stresses applied in the test. From the results from this load test, a design capacity of 178 kN (40 kips) per aggregate pier was approved.

Theoretical Uplift Capacity. The purpose of the following theoretical analysis is to provide a plausible explanation for the unusually high pullout strength and elastic behavior of the load test aggregate pier. Prior to performing this theoretical analysis, it is necessary to consider the changes in stress which occur in the matrix soil and the aggregate pier during the construction process. The in-situ soil is initially in an at-rest condition and if the ground surface is fairly flat, it is reasonable to assume that the major principal stresses are horizontal and vertical. After the cavity is excavated, the horizontal stress reduces to zero (with capillary suction keeping the hole open), while the vertical stress remains approximately constant, and it is reasonable to assume that the vertical face of the cavity is a principal plane because there is no applied shear stress. During construction of the aggregate pier, horizontal stresses are established along the aggregate pier-matrix soil interface.

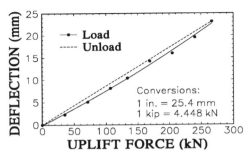

FIG. 3. Uplift Load Test Results from Meridian, Mississippi Hangar Project

Results from K_o-blade tests at other sites have indicated that full passive pressure can be achieved in the matrix soil (Lawton and Fox 1994). If it is assumed that no shear stresses are developed along the interface (so that the principal planes remain vertical and horizontal), and that full passive pressure is developed in the matrix soil, the interfacial horizontal stress can be estimated as $\sigma_h = \sigma_v \cdot K_p$, where $K_p = \tan^2(45° + \phi/2)$ (assuming $c = 0$); and ϕ = friction angle of the matrix soil. The stress state within the interfacial matrix soil is represented by Mohr's circle "a" in Fig. 4. As uplift force is applied to the aggregate pier, shear stresses develop along the vertical plane so that a rotation in the principal planes occurs and a major principal stress arch (Handy 1985) develops within the matrix soil as shown in Fig. 4. A similar stress arch would develop within the aggregate pier. Assuming that failure will occur along the interface and that the interfacial horizontal stress remains constant up to failure, the state of stress at failure would be represented by circle "b," and the interfacial shear stress at failure can be found from $\tau_{ff} = \sigma_h \cdot \tan \phi$. For this scenario to be correct, the interfacial vertical stress would have to increase from σ_{va} to σ_{vb}. Although it is reasonable to expect that some increase in interfacial vertical stress will occur as the vertical shear stress develops, it is not known if σ_{vb} is sustainable. If a lesser value of vertical stress is sustainable (σ_{vc}), a concomitant reduction in horizontal stress (σ_{hc}) would also occur, and τ_{ff} would be less than for the case of constant horizontal stress. Unpublished results from K_o-blade tests conducted by the third author adjacent to a model pile for the Talmadge Memorial Bridge in Savannah, Georgia may support the possibility of a reduction in horizontal stress during loading. The model pipe pile was driven open-ended, and was cleaned out and filled with concrete. The soil was at full passive pressure, apparently as a result of expansion when hit by sea water. As the pile was loaded in compression, it failed prematurely by plunging. As the load was increased, the horizontal stress decreased, explaining the 76 mm (3 in.) plunge to relieve the load. However, additional research and testing must be conducted to determine if the interfacial horizontal stress decreases when an aggregate pier is subjected to an uplift force, and what is the magnitude of the decrease (if any).

Using the previous discussion as the basis, the theoretical uplift capacity of the load test aggregate pier can calculated using the information shown in Fig. 5 and the following equation:

$$T_{max} = 0.5p[(\sigma_{h1} + \sigma_{h2}) \cdot H_1 \cdot \tan \phi_1 + (\sigma_{h3} + \sigma_{h4}) \cdot H_2 \cdot \tan \phi_2] + W \quad \ldots\ldots\ldots (4)$$

where p = length along horizontal perimeter of aggregate pier; σ_{h1} = interfacial horizontal stress at top of aggregate pier; σ_{h2} = interfacial horizontal stress in sand fill along boundary with clayey sand; σ_{h3} = interfacial horizontal stress in the clayey

130 DEEP SOIL IMPROVEMENT

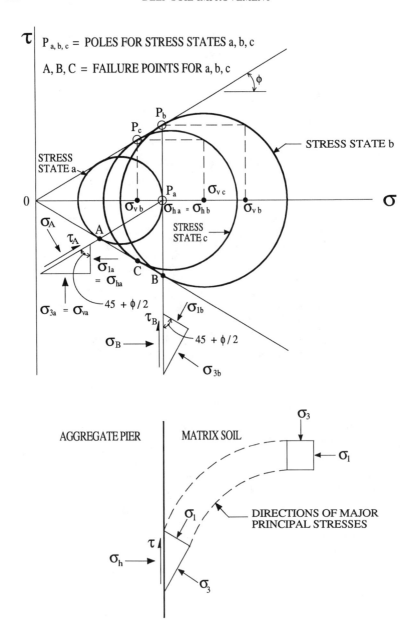

FIG. 4. Theoretical Stress States During Uplift of an Aggregate Pier

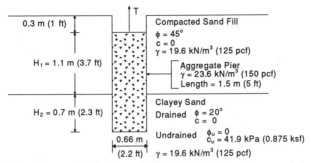

FIG. 5. Schematic Diagram of Uplift Load Test on Aggregate Pier

sand along boundary with sand fill; σ_{h4} = interfacial horizontal stress at bottom of aggregate pier; and W = weight of aggregate pier.

Assuming that full passive pressure was developed in the matrix soil during construction of the aggregate pier, and that the horizontal stress remained constant during the uplift, T_{max} is calculated as 663 kN (149 kips) short-term and 614 kN (138 kips) long-term. Both the short-term and long-term values are substantially greater than the maximum uplift force of 267 kN (60 kips) applied during the load test, and would be consistent with the small, essentially elastic deflections measured during the test. Whether these values overestimate or underestimate the actual maximum uplift capacity is not known since the uplift load test was not taken to failure. As discussed previously, it is possible that a reduction in horizontal stress may have occurred during the uplift process, resulting in a lower uplift capacity than predicted. However, two other factors tend to result in an underestimation of the uplift capacity: (1) The friction angles used in the analysis were obtained in the in-situ soils prior to installation of the aggregate piers. Borehole shear tests conducted on other aggregate pier projects have shown that installing an aggregate pier densifies the adjacent matrix soil, with a concomitant increase in shear strength; and (2) the actual dimensions of the aggregate pier were greater than nominal because of this densification.

As a hypothetical comparison, a similar theoretical analysis can be conducted to show that the uplift capacity of an aggregate pier should be substantially greater than the uplift capacity of a poured concrete anchor with the same dimensions. The key factor is the difference in horizontal stress developed along the pier-matrix interface during installation. Assuming the concrete is a viscous liquid, the horizontal pressure applied to the matrix soil is $\sigma_h = \gamma_{conc} \cdot z$, where γ_{conc} = unit weight of the wet concrete; and z = depth of concrete above the point being analyzed. This corresponds to a lateral coefficient for the concrete of K = 1 (based on γ_{conc}), compared to K_p = 5.83 for the sand fill and K_p = 2.04 for the clayey sand. If a stiff concrete mix is used, voids and irregularities along the interface with the matrix soil may result in lower horizontal stresses than predicted using this analysis. Using the same assumptions as for the aggregate pier and K = 1, the maximum uplift capacity for a poured concrete anchor with the same dimensions is 236 kN (53 kips) short-term (assuming an adhesion factor of 1.0 in the clayey sand) and 147 kN (33 kips) long-term. Thus, the theoretical uplift capacity of the aggregate pier is 2.8 times (short-term) and 4.2 times (long-term) that for the poured concrete anchor. Future full-scale uplift tests on aggregate piers and poured concrete anchors for the same subsoil conditions are planned to determine the validity of this theoretical procedure.

Performance of Aggregate Piers. Windstorms within the area of the site have been recorded to be as high as 97 to 113 km/hr (60 to 70 mi./hr) since the hangar structure was erected. No measurable uplift displacements have been recorded. Footing settlements were surveyed after erection of the structural steel and prior to the roof and door construction. No settlements were measurable with surveying instruments accurate to 0.25 mm (0.01 in.).

SUMMARY AND CONCLUSIONS

Two case histories in which short aggregate piers were used have been described. In the first project, an expansion of an existing hospital, aggregate pier-reinforced soils supporting both individual footings and mat foundations were effective in substantially reducing settlements. A method for estimating settlements of the aggregate pier-reinforced soils was described, with good correlation shown between predicted and actual settlements. Aggregate piers were used in the second project to provide uplift capacity for an airplane hangar susceptible to high uplift forces resulting from wind loads. The aggregate piers have performed well in winds up to 113 km/hr (70 mi./hr). Using a proposed theory, it was shown that the uplift capacity of aggregate piers is substantially greater than that of poured concrete anchors with the same dimensions. Future research and full-scale field testing is planned to establish the validity of the theoretical procedure.

ACKNOWLEDGMENTS

The authors wish to acknowledge the excellent cooperation of members of BEERS Construction Company, Inc., in providing information on performance of the Atlanta regional hospital project, and Tilley Constructors, Inc., for performance information on the Mississippi hangar project. The help and cooperation of Mr. James O'Kon and O'Kon Company, Inc. in providing background and design information on the hangar project is greatly appreciated.

APPENDIX - REFERENCES

Bowles, J. E. (1988). *Foundation Analysis and Design*, 4th ed., McGraw-Hill, New York, New York.

Burmister, D. M. (1958). "Evaluation of pavement systems of the WASHO road test by layered system methods." *Highway Research Board Bulletin 177*, 26-54.

Fox, N. S., and Lawton, E. C. (1993). "Short aggregate piers and method and apparatus for producing same." U.S. Patent No. 5,249,892 issued October 5.

Handy, R. L. (1985). "The Arch in Soil Arching." *J. Geotech. Engrg.*, ASCE, 111(3), March, 302-318.

Lawton, E. C., and Fox, N. S. (1994). "Settlement of structures supported on marginal or inadequate soils stiffened with short aggregate piers." *Geotechnical Special Publication No. 40: Vertical and Horizontal Deformations of Foundations and Embankments*, ASCE, 2, 962-974.

Schmertmann, J. H. (1970). "Static cone to compute static settlement over sand." *J. Soil Mech. and Found. Div.*, Proc. ASCE, 96(SM3), 1011-1043.

Schmertmann, J. H., Hartman, J. P., and Brown, P. R. (1978). "Improved strain influence factor diagrams." *J. Geotech. Engrg. Div.*, Proc. ASCE, 104(GT8), 1131-1135.

A DEEP SOIL MIX CUTOFF WALL AT LOCKINGTON DAM, OHIO

Andrew D. Walker[1], M. ASCE

Abstract

The Miami Conservancy District built five flood control dams after the catastrophic flood of 1913 in Dayton, Ohio. Lockington Dam was one of these five hydraulic fill structures. The Ohio State Department of Natural Resources (ODNR) issued new regulations in 1981 requiring that design flood for class I dams be considered equal to the Probable Maximum Flood (PMF). Flood routing studies performed for the PMF calculated inflow indicated that the reservoir rose to the crest of the dam. Therefore, the existing dam core required extending to the crest level to maintain dam safety. The extension was constructed using a deep soil mix soil cement cutoff wall. This paper will describe the cutoff wall from conceptual design through to construction, including alternates considered, construction methods and data obtained, both prior to and during construction.

While Deep Soil Mixing (DSM) has been used on contaminated sites as a method of containment and for dam foundations to decrease the risk of liquefaction, Lockington Dam represents one of the first applications as a conventional cutoff on an existing dam.

Introduction

The largest single project under construction in the United States in 1919 was the flood control works of the Miami Conservancy District.

---
[1] Project Manager, Geo-Con, Inc., 4075 Monroeville Boulevard, Monroeville, PA 15146

Started in 1918, the five flood control dams and 73 miles of levees which made up the project, would not be completed until 1921 (ENR, 1994).

Work was initiated in response to ten major floods which had inundated Ohio's Miami Valley at various times over the previous century, claiming more than 1000 lives and causing more than $100 million in property damage. Historically, this project saw the first ever use of electrical draglines on a large construction contract.

Power shovels excavated raw materials from the pits which were then mixed with water to form a slurry and pumped by dredge pipes to the dams. The water gradually drained and the dam rose in layers of impervious clay in the center, providing the waterproof core, with sand and gravel forming the dam shell. This use of hydraulic fill, while not a first, was one of the largest such applications ever recorded.

One of the five structures, the twenty-one meter high Lockington Dam, is located on Loramic Creek, a tributary of the Great Miami River, north of Piqua, Ohio. The 1950 meter long dam is founded on Cedarville limestone bedrock with a combined central low level outlet and emergency spillway (Figure 1). No permanent reservoir exists upstream of the dam as it is operated purely as a flood control project with storage only occurring when the upstream runoff exceeds the capacity of the outlet structure.

Figure 1. Aerial View of Lockington Dam

The central impervious core was built to about 4 meters below the dam crest which corresponded, at that time, to the maximum reservoir level for the Official Plan Flood (OPF).

Site Geology

Test pits were excavated in the summer of 1991 to augment boreholes sunk in 1983 as part of the original preliminary studies. This site investigation indicated that the general geological succession in the upper 6 meters of the dam consisted of:

- A medium dense grey moist gravelly sand to sandy gravel classifying as an SP to SM to a GW to GP material, 1.5 to 4.3 meters deep. Mostly gravel was encountered in the west side of the spillway with mainly sand encountered on the east side. As mentioned, this material was probably placed by pit run hydraulic methods. The absence of significant fines in this upper layer indicated that no attempt had been made to obtain finer grained fill for this upper zone. Based on the Unified Soil Classification, this material has a permeability ranging from 10 cm/sec (GW) to 3×10^{-3} cm/sec (SM). Falling head permeability tests in this strata indicated permeabilities as high as 4×10^{-2} cm/sec.

- A loose brown silty sand to soft sandy silt, wet to saturated in layers classifying as SW-SM to ML, 1 to 2 meters thick. This material appears to be the start of the core material. Test pits indicated a heterogeneous material, with thin layers of sand and gravel within the wet silt and fine sand layers. This layering is typical of hydraulic fill placement.

- A soft brown wet to saturated sandy silt to silty clay below the bottom of the above layer to the dam foundation, classifying as ML to CL, 1 to 8 meters thick. This material is the clay core and appeared to be somewhat finer than the layered strata above.

Design Considerations

Lockington Dam is classified as a Class 1 structure by the Ohio Department of Natural Resources (ODNR). Accordingly, under regulations issued in 1981, the design storm for the structure is considered equal to the Probable Maximum Flood (PMF).

Hydrologic studies performed in 1982, 1988 and 1990 indicated that the PMF would exceed the Official Plan Flood (OPF) producing reservoir storage levels higher than the design storage level. It was evident that the dam fill material overlying the core would transmit a significant amount of seepage under PMF conditions.

The client's consulting engineer, Harza of Chicago, estimated that during PMF reservoir storage conditions, the reservoir water level would be 1.2 meters below the crest for more than two days and 4.3 meters below the crest for more than one week. The resulting seepage through the upper core could produce potentially unstable conditions leading to a sliding or piping failure of the dam.

It was recommended that a cutoff be constructed in this permeable zone, extending at least one meter into the existing core material, thus assuring the safety of the dam during a PMF event.

A major consideration in cutoff selection was the potential for drying out and cracking of the wall with time. The upper part of the dam had remained above the highest reservoir level for over 70 years. With the sand and gravel having a moisture content of around 3.5 percent, a significant loss of moisture from the wall with time was anticipated.

A further requirement for the cutoff was an intimate contact at the spillway abutment walls that bisect the dam near its midpoint.

Cutoff Alternatives

A number of techniques were studied for forming the cutoff wall including:

- Open excavation and replacement with compacted clay material.
- Steel sheetpile cutoff.
- Pre-cast concrete panels.
- Soil-bentonite slurry wall.
- Jet grout curtain.
- Soil mixed cutoff wall.
- Cement-bentonite slurry wall.

Detailed costings indicated that the latter three methods were the most economical. The narrowness of the crest at twenty four feet, and the length of the dam, dictated remote mixing for a soil bentonite wall. A cement bentonite wall was therefore the favored slurry wall option. The slurry wall design incorporated a full depth HDPE liner to maintain wall integrity should possible desiccation of the backfill occur with time.

This was a genuine concern, given that the cement bentonite backfill consists of 90% water which is stabilized by the use of bentonite. As explained later, this is not a problem for a soil mixed wall, where the water solids ratio of the soil-cement-bentonite is low.

A major advantage of the soil mix wall technique was that there was no excavated soil for disposal, or a need for remote mixing operations, with all the mixing takes place *insitu*.

Accordingly, the bid package allowed contractors to bid either an auger mixed soil-cement wall and/or a conventional cement-bentonite slurry wall.

The price received from Geo-Con, Inc., for the soil mixed alternate was ten percent lower than the slurry wall and met all requirements of the specification, thus Geo-Con was awarded the contract. Interestingly, the price differential was very close to the cost of the supply and installation of the HDPE liner specified for the slurry wall.

Deep Soil Mixing

Pioneered in Europe and Japan, the Deep Soil Mixing System has made dramatic progress in the United States in the last few years with the completion of many successful projects involving a wide range of applications (Ryan and Jasperse 1989; Reams, Glover and Reardon 1989).

A specific American evolution of the technique has been site remediation of hazardous wastes where, for instance, high soil disposal costs are eliminated, nevertheless, the list of major structural walls and cutoff applications on clean sites continues to grow (Jasperse and Miller 1990). The Lockington site is, to the author's knowledge, the first use of Soil Mixing in the United States to raise the core of an existing dam.

Deep Soil Mixing equipment consists of a DSM rig and a grout plant. The base machine for the rig is a 175 tonne crawler crane fitted with a set of leads. The leads guide a series of four hydraulically driven, overlapping mixing paddles and auger flights (Figure 2).

The auger flights are 0.9 meters in diameter. As the hollow-stem augers penetrate a slurry is injected through their tips. The flights break the soil loose which the paddles then blend with the grout. As the augers advance to greater depth, the mixing paddles continue to mix the soils. When the design depth is reached, the mixing shaft rotation is reversed and the mixing process continues as the shafts are brought to the surface.

Figure 2. DSM Augers

An overlapping pattern of primary and secondary strokes is used ensuring a continuous wall, as at Lockington, or full coverage for block treatment (Figure 3). Depending on soil conditions, DSM can extend to over 30 meters in depth.

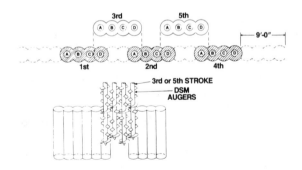

Figure 3. Overlapping Construction

The DSM technique is also used to construct earth retaining structures. Bending resistance in the wall is created by installing vertical H-Beams within the soilcrete columns immediately after mixing, normally at 1.4 m centers, thus forming a structural waterproof wall. This is an attractive solution for supporting deep excavations in urban areas especially when a high groundwater table is present.

Backfill Mix Design

The contract specification called for an in-place wall permeability of 1×10^{-6} cm/sec and a cement content of the soilcrete (soil and grout mixture) of not less than 6% by weight of soilcrete.

In addition, a mix design developed from bench scale testing was to be submitted prior to start of work to demonstrate that these properties were attainable. This procedure is standard for soil mixing projects where it is relatively straightforward to model in the laboratory the final wall composition using soil samples recovered from site.

An extensive laboratory test program was therefore developed with the following objectives:

1. Characterize the proposed materials.
2. Test and develop a workable cement-based grout suitable for soil mixing.
3. Develop and characterize trial grout-soil mixtures which span the expected range of material properties.
4. Perform unconfined compressive strength tests and hydraulic conductivity tests on the proposed grout-soil mixtures.
5. Recommend a proportioning of materials which will provide a grout-soil mixture with a maximum hydraulic conductivity of 1×10^{-6} cm/sec.

In total, 12 different grout soil mixtures were made during the program. Defining cement content by weight of soilcrete is impractical as soilcrete density varies depending on the soil bulk density and is only known once samples have been cast and density measured. On site the procedure is to add a certain volume of a fixed proportioned grout to each soil mixed shaft volume. This was the approach adopted in the laboratory with the grout to soil ratio by volume set at an empirical value which was the minimum required to allow satisfactory auger penetration.

The cement and bentonite contents of the grout were varied to assess the effect on the soilcrete properties, with cement/water ratios by weight ranging from 14% to 33%. The cement contents by soilcrete weight of each sample that resulted are given in Table 1 along with the test results.

SLURRY & GROUT PROPERTIES	1A	2A	3A	4A	5A	6A	1B	2B	3B	4B	5B	6B
Soil Sample	SP-SM	SP-SM	SP-SM	SP-SM	SP-SM	SP-SM	SP-SC	SP-SC	SP-SC	SP-SC	SP-SC	SP-SC
Grout Ratios C/W B/W G/S	33 4 32	33 4 37	25 6 31	19 6 30	19 6 34	14 7 29	33 4 32	33 4 37	25 6 31	19 6 30	19 6 34	14 7 29
VISC (CP @ 600 rpm)	99	103	140	153	167	186	99	103	140	153	167	186
UNIT WEIGHT (t/m³)	1.92	1.92	2.00	1.97	1.84	1.86		1.94		1.94		
USC(KN/m²) 3 days 7 days 28 days 60 days	136 358 411	150 353 406	140 212 276	66 75 103 330	94 138 241	53 57 44 121	132 76 576	132 240 92 217	138 203 220	96 129 191	114 136 178	98 82 147
CEMENT CONTENT % BY WT. OF SOILCRETE	5.9	6.5	4.5	3.5	3.9	2.6		6.5		3.5		
PERMEABILITY (cm/sec) x 10⁻⁴	.091	.195	0.290	.310	.691	.192	-	1.02	-	.175	-	-

Table 1. Trial Soil-Grout Mix Proportions and Properties

Unconfined compressive strengths were obtained as per ASTM D1633 while hydraulic conductivity testing was carried out using triaxial (flexible wall) permeameters as per ASTM D-5084. Tests were performed using an average hydraulic gradient of 15 and a confining stress of 345 kN/m² to model actual wall service loads. Samples were moist cured for durations between 14 and 21 days prior to testing.

All samples tested met the specified permeability of 1×10^{-6} cm/sec and, as anticipated, a general increase in strength was observed with increased cement/water ratio.

Normally for cutoff barriers the residual permeability is the primary consideration; but for the Lockington project, the cement content was determined to be of equal importance. It was felt that a grout of higher cement content would provide greater erosion resistance should overtopping of the dam occur. Conversely, from past experience, it was known that increasing cement content could compromise permeability, probably due to the more open lattice structure produced by cement hydrations having a higher permeability than that of bentonite and soil. This correlation has been demonstrated clearly in the past for neat cement grouts (Littlejohn 1982) but due to the present small database, has not as yet been clearly proven for S-C-B mixes.

Desiccation tests were carried out on samples 1A and 4A, which appeared the most favorable from the initial tests. Samples were set in beakers surrounded by moist gravel and left for two weeks without humidity control. Moisture contents reduced fourfold to between 7% and 8% with no sample deterioration observed. As a more extreme test, the specimens were oven dried, and again showed no deterioration, the samples remaining monolithic with no visible cracks.

Given the emphasis placed on strength, the high strength Mix 1A was ultimately selected as it also met the other requirements of the specification.

Construction

Construction took place over a seven week period during the harsh winter of 1993, with work often taking place in sub-zero temperatures. Nevertheless, the work proceeded as planned with only a few days lost to weather with production peaking at 650 m^2 per shift. A total of 6,200 m^2 of wall was installed down to a maximum depth of 6.5 m. No penetration problems were experienced by the drill rig which had a rated output of 67,000 Nm torque, developed from a 450 kw powerpack; and excellent mixing was achieved.

The first work on site was to construct a 1 m wide by 1 m deep trench on the wall centerline to act as a containment for the residual material displaced during mixing. This quantity was around 20% of the wall volume.

In order to overcome the narrowness and length of the site, the grout plant was erected just east of the central spillway for construction on the west side and then re-erected on the east abutment for mixing on the east side. The restricted width of the spillway bridge also necessitated teardown and re-erection of the crane after completion of the west section of the work. Even so, pumping distances for the grout were over 900 m which was accomplished by the use of four Moyno L-10 grout supply pumps, one for each auger.

The crest was just of sufficient width for the Manitowoc 4100 crane on which the drill rig and powerpack were mounted, with the complete unit supported on crane mats to spread the high equipment load (Figure 4).

The block treatment at the spillway wall to provide a watertight connection was built using the jet grout system (single jet). Before the last soil mix shaft adjacent to the wall had reached its final set, the jet grout monitor was inserted to the base of the block using a conventional hydraulic-driven tracked drill rig. Jet grout columns were then formed to connect the soil mix wall to the concrete face of the wall. The grout jet at a pressure of 20.7 MN/m^2, scoured the concrete face and simultaneously cut into the last soil mix column as it was rotated. The same grout mix as for the DSM work was used in the jetting with the monitor raised at 1.5 m per minute.

Figure 4. DSM Unit in Operation

This method produced a much better seal than could have been achieved by some form of mechanical cleaning of the wall followed by column remixing.

Quality Control

The specification required a very thorough QC program to include procedures to check the verticality, depth, and alignment of the wall, and the grout proportions and application rates, to ensure the highest standards of workmanship.

The verticality and alignment of the wall was achieved by two controls. A laser provided a line onto a target on the DSM auger shaft for horizontal alignment, while verticality was monitored by two measurements made on the mast. These were the pitch and the roll. Two servo accelerometers within a fixture on the leads continuously outputted the angle to a display in the operator's cab.

At the grout station, bentonite slurry and cement slurry were mixed separately before being mixed to produce the final bentonite-cement slurry in a 3.8 m³ lightning mixer. Slurry densities and viscosities were taken by mud balance and marsh cone respectively (see Table 2) to monitor batch control.

The amount of grout injected in each DSM stroke was controlled by the DSM technician using automatic flowmeters, one for each grout pump. The grout flowmeters also allowed adjustment of pump speeds to be made between individual augers. This is helpful when redrilling previously drilled columns for overlapping, or to aid penetration in difficult areas. It was soon found at Lockington that these features were not needed due to the very consistent rate of penetration and the system was changed to one single-instrumented L-12 pump.

SUBJECT		STANDARD	TYPE OF TEST	MINIMUM FREQUENCY	SPECIFIED VALUES
MATERIALS	Water	EPA Standard Methods	- pH - Total Hardness	Per water source or as changes occur	As required to properly hydrate bentonite with approved additives
	Thinners	-	Manufacturer certification	One per truckload	As approved by Engineer
	Bentonite	API STD 13A	Manufacturer certificate of compliance	Truck load	Premium grade sodium cation Montmorillonite
	Cement	ASTM C 150	Manufacturer certificate of compliance	One per lot	Portland, Type 1
BENTONITE SLURRY	Prior to Addition of Cement	API STD 13B	- Viscosity - Density	4 per shift 4 per shift	MFV ≥ 35 sec. Density ≥ 64 pcf
C-B SLURRY	At Mixer or Surface in the Trench	API STD 13B	- C/W Ratio - Density	- Continuously - 4 per shift	C/W ≥ 0.19 70 ≤ Density ≤ 90 pcf
SOIL-CEMENT BENTONITE	Grab Sample	ASTM D 1633 ASTM D 5084	- UCS - Permeability testing with hydraulic gradient of 15	1 per shift 1 per shift	k ≤ 1 x 10⁻⁴ cm/sec when fully cured

Table 2. Materials Quality Control Program

In addition, extensive sampling of the wall was required to verify compliance with the specification. The uncured mixed-in-place column was sampled using a specifically designed tube, following completion of a specific DSM stroke. The tube was closed by a pneumatic device at the designated depth and the sample brought to the surface.

Samples were taken every shift for permeability and unconfined compressive strength testing. The results of the tests are shown in Table 3.

The primary intent of obtaining higher strengths was achieved, with a few slightly lower compressive strengths in thick zones of ML and CL material where possibly no vertical blending with coarser soils took place. Even so, these lower values were comparable to typical cement-bentonite slurry wall strengths. As predicted, some permeability values were slightly higher than the 1×10^{-6} cm/sec which was originally targeted.

Furthermore, the soilcrete strength will continue to increase over an extended time period with the gain in strength from 4 days to 28 days of around 100% for this project reflecting previous experience which has recorded strengths still increasing up to 5 months after mixing (Ryan and Jasperse 1989).

SAMPLE NO.	SAMPLE DEPTH (m)	γ_d (t/m^3)	PERMEABILITY (cm/sec)	UNCONFINED COMPRESSIVE STRENGTH (KN/m^2)
D1	2.1	1.41	9.31×10^{-7}	73 (7 day)
D2	2.1	1.22	1.63×10^{-6}	229 (7days)
D3	2.1	1.30	1.29×10^{-6}	124 (7 days)
D4	4.6	1.39	8.48×10^{-7}	136 (7 days) / 334 (36 days)
D5	2.1	1.26	3.48×10^{-6} / 3.09×10^{-6} (Duplicate)	170 (7 days) / 423 (35 days)
D6	5.2	1.47	1.63×10^{-6} / 9.89×10^{-7} (Duplicate)	157 (7 days) / 354 (34 days)
D7	5.2	1.43	1.82×10^{-6}	114 (7 days) / 294 (33 days)
D8	2.7	1.40	1.12×10^{-7}	170 (9 days)
D9	1.5	1.03	5.28×10^{-7}	210 (7 days)
D10	1.5	1.51	5.4×10^{-8}	194 (7 days)
D11	1.8	1.40	1.1×10^{-7}	226 (7 days)
D12	1.5	1.55	1.05×10^{-7}	38 (14 days) / 28 (28 days)
D13	1.5	1.44	1.07×10^{-7}	75 (7 days) / 113 (28 days)
D14	1.5	1.51	1.83×10^{-7}	332 (7 days)
D15	1.5	1.42	8.96×10^{-8}	264 (7 days)
D16	3.7	1.73	1.48×10^{-7}	85 (7 days) / 127 (28 days)
D17	3.7	1.50	5.32×10^{-7}	170 (7 days) / 233 (28 days)
D18	5.5	1.63	1.27×10^{-6}	122 (7 days) / 189 (28 days)
D19	3.7	1.49	3.7×10^{-7}	43 (7 days) / 65 (28 days)
D20	3.7	1.48	1.36×10^{-6}	69 (7 days)
D21	3.1	1.48	2.04×10^{-6}	157 (7 days)
D22	2.4	1.59	2.01×10^{-6}	360 (7 days)
D23	1.8	1.31	1.35×10^{-7}	332 (7 days)

Table 3. Wall Permeability and Strength Results

As a final measure, a 5 m by 5 m test panel was excavated to investigate the degree of mixing. The wall exposed was continuous and well blended having the consistency of a firm clay, after a two-week cure period.

Closing

As the older water retaining structures in the United States continues to age and deteriorate, there will be a growing need to remediate and upgrade these structures to meet new regulations and factors of safety.

The project at Lockington is a good example of a recently introduced innovative technology, namely Deep Soil Mixing, providing the best solution both technically and economically, to the specific problems of an existing dam.

DSM was the only method that could provide a low permeability wall with good resistance to erosion and desiccation, all verifiable properties of a soil-cement-bentonite cutoff, and still overcome difficult physical site constraints, through mixing all materials *insitu*.

Furthermore, the use of preconstruction laboratory testing accurately predicted the *insitu* properties of the wall which were subsequently confirmed by an extensive post-construction testing phase.

Appendix   REFERENCES

ENR, January 3, 1994.

Jasperse, B. M. and Miller, D. A., "Installation of Vertical Barriers Using Deep Soil Mixing", Hazmat Central '90, Rosemont, Illinois, 1990.

Littlejohn, Dr. G. S., Design of Cement-Based Grouts, Conference on Grouting in Geotechnical Engineering, New Orleans, Louisiana, 1982.

Reams, D., Glover, J., and Reardon, J., Deep Soil Mixing Shoring System to Construct a 60 mgd, 40-ft Deep Wastewater Pumping Station, KOR Engineering, 1989.

Ryan, C. R. and Jasperse, B. M., "Deep Soil Mixing of the Jackson Lake Dam", ASCE Geotechnical and Construction Division, Special Conference, June, 1989.

Subject Index
Page number refers to first page of paper

Aggregates, 121
Alluvium, 80

Budgets, 68
Buildings, 55

Case reports, 55, 80
Coastal environment, 90
Collapsible soils, 26
Construction, 68
Containers, 44
Core walls, 133
Cost minimization, 55

Dams, 133
Dams, embankment, 1
Displacement, 44

Earthwork, 68
Embankment stability, 1
Excavation, 106

Fills, 68, 90
Finegrained soils, 80, 90
Flood control, 133
Foundations, 121

Granular materials, 121
Ground-water management, 106

Hydraulic fill, 90, 133

Landfills, 26, 55
Liquefaction, 1

Mines, 26
Mixing, 106, 133

Offshore engineering, 90
Ohio, 133

Peat, 26
Piers, 121
Powerplants, 55

Quality control, 26

Radioactive waste disposal, 44

Sand, 68, 80
Seepage control, 106
Seismic stability, 1
Settlement control, 121
Silts, 80
Silty soils, 68
Soil cement, 133
Soil compaction, 1, 26, 44, 55, 68, 80, 121
Soil consolidation tests, 90
Soil stabilization, 55
Soils, 106, 133
Soils, saturated, 80
South Carolina, 44
Stone columns, 121

Testing, 1
Tunnel construction, 106

Vacuum, 90

Walls, 106
Waste disposal, 26, 44
Wick drains, 1
Winter, 68

Author Index
Page number refers to the first page of paper

Bayuk, Albert A., 55
Beaton, Nelson F., 80

Dise, Karl, 1
Dumas, Jean C., 80

Fox, Nathaniel S., 121

Handy, Richard L., 121

Jacob, A., 90
Johnsen, Lawrence F., 68
Juran, I., 90

Kavazanjian, E., Jr., 90
Kim, Ji-Hyoung, 26

Lawton, Evert C., 121

McMullin, Scott R., 44
Morel, Jean-François, 80

Rollins, Kyle M., 26

Stevens, Michael G., 1

Takeshima, Shigeru, 106
Thevanayagam, S., 90
Tonzi, Christopher, 68

Von Thun, J. Lawrence, 1

Walker, Andrew D., 55, 133

Yang, David S., 106